Big Snow, Little Snow

Big Snow, Little Snow

by

Trudy Ann Parker

Dawnland Publications
P.O. Box 223
Lancaster, NH 03584

ISBN 0-9662523-2-2

Typeset by CDR Computer Services, Wilmington, VT
Printed at Commonwealth Priniting, Hadley MA

1 2 3 4 5 6 7 8 9 10 99 00 01 02 03 04 05

Big Snow,
Little Snow

Log Driving on the formidable Connecticut River was adventurous and deadly. That river was as wild and unrelenting as the Log Drivers who challenged it and on its back moved the mountains to the sea. The Abenaki Indians named it, the "Gwenitegok" meaning the Long Tidal River. We know it as the "Connecticut River" and it remains today unsurpassed in beauty and unequaled for its historical contribution to the heritage and shaping of New England.

– *Trudy Ann Parker*

For my Grandfathers John W. Call and Joseph Pinette

Both
Connecticut Valley Log Drivers

"Gwenitegok niona gwasitozik gilow ni gia zibo ginobak."
Long Tidal River we honor you and your brave rivermen.

Connecticut Valley Lumber Stamp

Forward

The tall trees, the roiling waters of the Connecticut River and the giant sawmills at Mount Tom beckoned – and a rugged breed of men and women heeded the call to get the wood out and let "daylight in the swamp."

This is their story, a story of hard work and hard living, a harsh yet beautiful land, and a way of life that lasted only as long as the tall trees – and then like yesterday's ripple on the river, was gone.

– *John D. Harrigan*

Preface

The old ones said "It was the time of trees and men." The fate of one depended on the other. The tree that had always been the provider of oxygen was now in great demand, the tree I mean and not the oxygen. That threat would come much later.

It was a time when "Men were Men," and Womean were said to be glad that they were. This United States of America was young, young and spirited, bursting at the seams to show the rest of the world that there was no civilized country anywhere on earth like America. There was no people more proud or more productive than was that industrious and determined breed called "Yankees".

This was also the era in which the great spruce forests of northern New England would be moved to the sea. There was only one way to do it and that was by water and this was when a tough and burly group of men shouldered their axes, crosscut and bucksaws and took to the woods. Isn't it strange that in order to build a country up you first had to cut it down?

Textile mills dotted the landscape in the north and south, proving that cotton was king, and the economy flourished.

The steamboat came to haul the cotton to a rich market

and there were southern cotton barons but in the northern regions of the Connecticut River Valley there were men called land barons or lumber kings. But, this is not their story. This is the tale of magnificent men and magnificent trees.

Timber! This was the cry that echoed through the massive northern forests, because the tree was the source from which lumber originated. A growing nation needed lumber for its homes and factories and their furnishings.

Another item was made from trees a little later on and that was paper. Paper would be in great demand here and abroad as the business of every nation was recorded on it.

And so they came, the men and boys from every northern hamlet and lived in crude logging camps all winter and cut logs on days that their noses and hands didn't freeze doing that chore. Then the logs were dragged to landings near streams and rivers so that come spring, they were rolled into the water to be guided or driven to the sawmill at Mt. Tom, Massachusetts.

The long ago logs floating along the Connecticut River came from the direction of the four winds. From a thousand brooks and streams that emptied into this river they came. When the ice went out, and the water was just right, sluiced logs came to join the drive from the Ammonoosuc River, Nulhegan River, Indian Stream, and from many others too numerous to name. Some of the logs came from the Swift Diamond into the Magalloway, then into the Androscoggin, through the Thirteen-Mile Woods to Berlin; then westward to the White River to take the logs on to the Connecticut. It was said that the Androscoggin Drive actually floated more timber than the Connecticut Valley loggers did, but they held up at Woodsville, N.H. or Wells River, VT., because they didn't want the hassle of confronting old George Van Dyke and his river hogs, who would let no one go ahead of them, no one.

From the headwaters at Third Connecticut Lake to Bellows Falls, there is little evidence today that thousands of men once filled this river with uncounted millions of logs and rode them sharpshod to the mills below. It looks to be a quiet stream. No longer do the banks echo with sounds of working men in red flannel or woolen shirts.

The Connecticut Valley Log Drive came down the river for half a century with all the excitement and fanfare of a circus coming into town, and by the time they arrived, only the dead didn't know they were there.

Their white tents set up in neat rows along the Upper Coos Meadows always forewarned that soon the meadow and the town would be teaming with life.

Then just as suddenly, the men, the horses, the bateaux, and of course, the logs were gone, and there was a great solitude that prevailed now on the streets of North Stratford and every other river town in the Connecticut Valley. On the New Hampshire side of the Connecticut River, business was always better because those towns sold liquor. Vermont was a dry state and New Hampshire was where dry men sated their thirst, so they said.

With the passing of the annual drive, the verdant meadows lay bare of bellowing men and the thunder of floating timber. The tents were gone and the flickering night fires of the men could no longer be seen. In one brief time it had happened in an era like no other in the annals of New England history. The Drive had "gone down" but like the pussy willow and the robin a Drive would pass this way again, come spring. Another year, another drive, another spring.

Spring was high water and life in the air and winter was slowly letting go of its terrible grip on the northern landscape. Young women yearned for the fragrance of flowers and a warm spring sun. Young men, most able bodied ones that is, were gearing up for the ice to go out, so the spring log drive could begin. It was a time when little boys dreamed of being big men, big, river-driving men. With enough education and study doctors and lawyers could be made, but not those undisputed "Lords of the Wilderness." Log drivers, who felled timber in the fall and winter months and then rode those mighty logs to market when the log drive started in the spring, now there was the measure of a man! No amount of book learning could make a river driver. He had to be born to it or else it was in his blood. If a man was clumsy or had slow reflexes, then that old river got him. If a man could beat the mountains and on top of that could beat the river, it was understood that he could beat anything else that life might confront him with. The Connecticut River was a mighty force that taught men hard, but taught them well.

For every little boy who dreamed of wearing a top hat and carrying a cane, like a fine gentleman, hundreds dreamed of wearing spiked boots. There was a place for gentlemen back then, but not in most boys' hearts. When the Connecticut Valley Log Drive snaked its way down the valley, even the one-room schools often recessed early so the boys and girls could see this sight that occurred every spring. As a matter of fact, this was when spring began in this valley. There would be no spring without the drive. And for almost fifty years there would be a drive, come Hell or high water, big snow or little snow, if there was a spring, the Drive was going down.

Wistful boys watching from nearby meadows and bridges as the drive "went down" or progressed through the lush, green valley would be the next generation of log drivers, at least that was their dream. After all, what man in his right mind didn't desire to be one of those hard-drinking, hard-fighting icons of the valley around whom legends were built? Some men of course, but to a boy's way of thinking, those log driving legends were as fearless as Attila the Hun and as graceful as a ballet dancer. What a sight to see, those magnificent men, with pike poles in hand flitting gracefully from log to log. How easy they made this job look with their sharp eyes always searching for the starting of a possible log jam up ahead. Sometimes those roaring river men would do a little dance upon a log for the benefit of their captive audience. These were the undisputed "Lords of the Wilderness," that rare breed that let daylight into the swamps, moved the mountains and walked upon the water.

There was then or is now no better season nor place than spring in the northern reaches of the Connecticut River

Valley. Robin redbreast is still the first sign of spring, but long ago, in the uppermost part of this valley, spring began when the annual log drive commenced. This drive started at First Connecticut Lake in remote Pittsburg, N.H., heading down the Connecticut River to Mt. Tom, Massachusetts where it ended. From many different streams other drives would join in. This was two hundred and forty miles of wild and treacherous river and millions of feet of floating logs. What a sight it was, this scene of timber without end as the dark spruce forests of northern New England began the deadly drama of their journey toward the sea at Long Island Sound, N.Y.

An army of men, many only boys, rode the rapids every spring after spending cold, northern winters in drafty logging camps. Cutting, sawing and piling timber was only part of the story. Moving the mountains to market was the other half of the experience they called river or log driving.

When a man signed on as a log driver he did so knowing that he would be paid only when the logs had reached their destination in the spring. If any man chose to leave the job before that, he left unpaid. Realizing fully the consequences of quitting early, few good river men ever quit. They were a tough, determined breed, who liked what they did. No, that is an understatement, they loved what they did, and they made sure no one did it better!

Often in the course of the annual drive, many logs did not make it, and in some cases it was some of the men who didn't make it. Many a man lost his life trying to prevent or free-up a log jam and often his fate was sealed by the dark and swirling water of the Connecticut, or the streams that ran into it. "If you live by the sword, you shall die by the sword," goes the quote and often the only epitaph or grave marker for the log driver or "river hog" was a pair of spiked boots hanging from a handy tree to tell the world one river man's fate. And it did.

If an epitaph was ever written to a log driver it might say that he was a woodsman, a reputed expert with a peavey, an ax, or a bucksaw. It might say he was a teamster, a bateau man, a cook or a scaler. Some of these lumberjacks were great teamsters, and others had great expertise in explosives and the art of demolition. There were camp clerks, blacksmiths and harness boys, because log driving had it all. There was a great need for many kinds of knowledge and skill and all of these were used in the business of bringing down the drive.

Many of the old lumber camps were nearly all roof and some were only two or three logs high with small poles flattened down on one side that served as a quaint flooring over the ground. Heat and cooking were done on a fire pit which took up most of the floor and ventilation was provided by a smoke hole directly above the fire pit. The earliest of these lumber camps had one huge bed, with quilts up to and more than twenty-five feet. Ten to twelve men slept in that communal bed, lying on their sides (not on their backs) much like spoons. It was said that when one turned, all the men did the same and they called this "flipping out." And, that would "flip quite a few of us out," but the long ago river driving man stayed with the program. He knew these rough but sturdy hovels were not designed for the comfort of the loggers, nor for their health and welfare, but to give the men basic shelter from the elements and enough heat, hot food and sufficient ventilation for these tough human beings to survive.

When the first stoves came to the lumber camps they were welcomed and appreciated because a man could get both sides of him warm.

If a river man was unfortunate enough to get his spiked boots wet, he wore them until they dried, dried on his feet. If you don't think ten men could sleep on their sides like spoons, many wearing wet spiked boots, you are wrong. They did. And if we had worked from first light until dark, bucking wood, sleep would be welcomed. There were no sleeping pills, exhaustion did nicely.

In later years bunks were provided, and on these the men with the wet spiked boots could sleep by themselves on their sides, stomachs or backs, in a deep hypnotic sleep as their boots dried and exhaustion was again the tonic which mercifully carried them to a much needed rest.

Clothes lines were strung across the camp and when there was time the men could wash their own socks and underwear and hang them up to dry. And speaking of staying dry, that was always foremost on the mind of a river man who was wet most of the time. It went with the job, and the smell of drying socks was not very pleasant but you combine it with the odor of "mutton tallow" (sheep fat) and you really had a rank smell. This too went with the job. Washing the socks was optional but most of the men covered their socks with this lard or fat because it helped protect them against water. This was done before they put on their most precious possession ..... their spiked or caulked boots. Survival, that was the name of the game and so often only those spiked boots stood between a man and his death on the Connecticut River.

Most sizable logging operations had "cookees" to help the cooks and stable boys to help the teamsters, and if they were lucky they had a good hunter among them who could go out and bring in some fresh game. This game and every other staple was handed over to some of the world's most renowned cooks, the logging camp cooks. Theirs was a cuisine void of marinades and sauces, except for browned flour or milk gravy. These gravies were served over whatever the cooks baked, boiled or fried. The only marinades made were for pickles and they were said to be the best. There was an unwritten code that any meat or chicken that had to be liquored up to come to the table wasn't fit to eat. This was understood in logging-camp cooking and what was expected was fluffy biscuits as big as the palm of a man's hand and cookies the size of dinner plates! A good cook was the greatest asset to a camp because a good cook meant a happy camp and happy campers were productive.

These burley, industrious men were not fussy eaters, but they demanded huge servings and simple but clean food. They also liked enough light when they ate, so that they could see if they were eating flies or blueberries. Some of course boasted "they could eat a peck of dirt," but most of the men would enjoy it more if they didn't know it.

Most logging operations fed the men very well, not

fancy but healthy, good food. And it was often said they "fed 'em like kings and worked 'em like horses," but that was all right too, because the very breed of log driver felt close to the earth, and the river drivers of Connecticut Valley Logging referred to themselves as "River Hogs."

In the era when log drives flourished so did the economy. There was money to spend. Working people are also spending creatures, and still there was frugality in these times of mild prosperity, as most local banks would loan money to trusted borrowers who in turn invested this capital into new businesses or to buy more land, wilderness, houses, farms and sawmills. All these things created work. The fairly good economy thus enabled an enterprising man of purpose to use the economy to his advantage, and this was done frequently and a nation prospered and grew. And in such times as these, were there ever hard ones? Yes, but like the northern mountains that protect her people from fierce winds, the Connecticut River Valley protected, sheltered, and cared for her own; providing men with an income, insuring an ambitious people employment, food, shelter, comfort, and confidence in themselves and in each other.

Can the productivity of a nation be measured by the pride, individual talent, and ingenuity of its workforce? Yes, it most surely can and did, in fact, during those best and worst of times in the days when great logs and pulpwood thundered through the valley, and no mortal man nor cotton either was the king. Timber was King.

The world at that time in the Connecticut River Valley revolved around timber and at such a pace it went. In the timber business or logging industry there was always work for a man or a boy with an ax or a bucksaw. Experience always helped, but if an able appearing youth seemed reasonably responsible and willing to work, the numerous lumber camps that dotted the landscape would usually hire him on the spot. No man was worthless. If a boy wasn't lazy he could be taught to be productive somewhere. This was the philosophy of the era and it worked. The forests of northern New England would always provide a livelihood for a young man with stamina and sand and a lore of the wilderness.

The rule applied that everybody could be taught to do something useful, to be an asset. If you were young what could be any easier than falling off a log? But, if you did that too many times, perhaps you would be better suited for a cookee (cook's helper). A log drive had many and various jobs and most men could be taught to do one of them.

Many farm boys, because they were used to animals, became teamsters and teamsters' helpers. You always started from the bottom, but you could achieve your goal job. It was up to the man. Success or failure were entirely up to him.

Can you swim? This question was of course frequently asked and nearly ninety-five percent of all the men working in northern logging operations stated that they could. In actuality, it was believed that only about forty-five percent of that number could, but when you can walk on the water who needs to swim?

From the virgin spruce forests of New England came a select few called Lumber Barons, and this often self-made breed came to possess great power and wealth. These were the "Lumber Kings" of New England and some lived to retire in luxury and garnered the lion's share of the profit reaped from the New England tree harvests that floated annually to market. Few ever forgot the dissipated face of the self-made George Van Dyke of Connecticut Valley Lumber, whose determination, intelligence, and genius moved the men who moved the mountains. Never forgotten either were his petty weaknesses which made him a lesser man than he was, like in life and death situations on the drive, when he would bellow like a bull, "To Hell with the man, save the cant dog!"

On a cold March day I stood atop the seventy-five foot cliff (called Riverside) overlooking the dam at Turners Falls in Montague, Massachusetts, as spring sleet and a raw wind whipped my face. I thought of George Van Dyke. This was the ledge where Van Dyke and his driver "Shorty" Hodgdon, seated in Van Dyke's big red automobile had rolled down that steep ledge toward the Connecticut River waiting there below.

This happened ninety years ago, in August of 1909, the very year Van Dyke and his river men had brought fifty-three million feet of timber down the Connecticut River. Van Dyke and Shorty, his chauffeur both died later of the injuries they received in that plunge.

That old lumber king who so many times had beaten the Connecticut had failed today, because this man who lived his life so close to that river would die of injuries received as he landed on the dry bottom of it. That fateful day the river won.

I thought too about Van Dyke's rivermen who struggled to get the timber baron's driver out of the wreckage on the rocks below and to administer care to their leader, the mortally injured Van Dyke. He had sensed the danger and had jumped out of the car. He then rolled down the cliff, landing beside the overturned auto. And here we see the best and worst of men. As some of the rivermen on the dry river bottom below strained and pulled to free the trapped chauffeur, others out sluicing on the dam who had witnessed the accident chanted, "to Hell with the man, save the matches!" The early morning sun came around and so did what goes around, it seems.

I thought of that complex, admired and even hated lumber king and of all those men working with their peaveys and pike poles below sluicing the logs through the dam. The churning water was strewn with chunks of unmelted ice and snow and I shivered as I thought about them.

What kind of men were these? Well, they were special, very special indeed, those "Cowboys of the Connecticut." They were an extension of the earth and a part of this eastern environment where endless streams and rivers came rushing down from the tangled, spruce-strewn mountains that they called home. Lumberjacks felled the New England forests, but it was the River men who rode those logs to market, armed with confidence, their trusted pike poles and magnificent grace. It was, however, that fierce pride they took in themselves, their work and in each other, that made them the fearless legends that they are. Come Hell or high water they took the drive down. This is the story of how they did that.

It is about men like both of my grandfathers and perhaps yours. In this novel I have strived to bring alive a sample of the history they made. The very heart and life-blood of the long ago log drives was the men, those arrogant, and unsung "Cowboys of the Connecticut," their dreams, their courage and their very own legacy. They were by all civilized or uncivilized standards not just survivors, they were as magnificent as the trees they rode to market. And as we follow this drive down, we will not see super men, just mortal men who did super human things every day, every phase of the drive.

Virgin timber was felled or sawed on the frozen winter tundra and drawn by sleds to the banks of streams to sit and wait for the annual drive. Getting the timber to the landings and ready to float in the spring was also part of the drive. The drive was complicated, and it took courage and skill and hundreds of good men doing many different jobs to bring the drive down. And there was usually not a man of them who had a job that was considered of little importance, nor who didn't take his part in the whole log driving picture very seriously. Whatever they did, they gave it their all. They were minute men ready to defy the elements, dance on a log or die around the next bend in the river, if that's what it took to get the job done. And they loved this dangerous occupation because they were part of the earth and yes, the earth was part of them.

These men lived close to the earth, felt its heartbeat under their wet feet, tasted its savory flavor in their mouths by day and ended each day smelling of sweat, and pitch and spruce. And by night, in the remote settings of those lonely logging camps, these lumber jacks who swung their axes all the day could hear the silence of the surrounding forest, the cry of the catbird, and the warning of the loon and the owl. And they loved these things just as they did the forest itself. This is why they did not tear it all to Hell, and they did not deplete it, they only took a part of it, and moved it to the sea.

The longest log drive in New England began with logging camps big and little that freckled the landscape in the Connecticut River Valley. Here like every other harvest of nature, each season has a purpose, a thing that must be done right then and there, and so it was with the log drives.

Those river rogues, or hearty log drivers of long ago, did each phase of their work with pride and did it as if their lives depended upon it. And they did. It was the only way that a man could beat the Connecticut.

This log drive was the last New England battle field, a place for ambitious young men to prove themselves, and they did, these young town and farm boys of English, Irish, Scottish, German, Polish, or whatever descent. A credit to their race also were the fearless, agile Frenchmen who walked the logs sure-footed and boisterous. And we must not forget the graceful magnificence with which the Abenaki, Mohawk and Mic-Mac Indians could guide a canoe or a bateau in the wrath of raging waters. Even on Connecticut River waters the courage and skill of the bateau men of the Bangor Tigers was marveled and respected. These men were Old Town Penobscots and Abenaki Indians, however your race didn't really matter here, only your job mattered, your life mattered, as these things were entwined. You see, the race that counted on a log drive was the one to save your life, yours and that of the fellow beside you, set against the turbulent high water of the Connecticut River. With so many men's lives depending on one another, skin color was never a factor, but courage, skill, and dependability were. A man's faith in himself and in his comrades was what mattered most.

At the end of a drive the Connecticut Valley River Hogs usually got into a damned good brawl with the Bangor Tigers over which outfit was best, but while the drive was going down, this petty business was put aside. There was but one goal, get the logs to market, as many as could be sent there, the most inexpensive way. On a drive, log drivers from different outfits worked together frequently. They might fight after they reached their destination but not while getting the drive down there. They worked together well, pride of outfit put aside. It was a Herculean challenge, conquering the Connecticut, and it took one damned tough group of river men acting as one unit to make the trip successful. This river was a dreaded enemy. It took a mighty army to beat it.

On lazy summer afternoons, riding along and gazing at this listless, scenic river, it is hard to imagine that this placid ribbon of silver and blue (called by early historians, "The River of the Sokokis," one of the Abenaki groups) was the highway to the future in the lives of those unsung heroes of the past. Headlong into the swirling water and history books, they rode the logs and defied the dreaded rapids.

The King of England, who the Hell was he? Few of those long ago log drivers could have told you his name but they knew "Who" the king or key log was, and what they had to do to dethrone that lethal monarch. And getting that one wasn't an easy chore, it was often a life and death situation. Clothed in wet bark, slippery and mean, he was one hell of a tough and determined adversary. But clothed in spiked boots and in wet Johnson pants, and armed with a pike pole or a cant dog, so was an agile north country river man.

Speaking of the King of England, who were these river drivers? And, where did they come from? Well, many came from Canada on foot, immigrants like my maternal Grandfather Joseph Pinette, a French river man. Others like my paternal Grandfather John Call, a Native American and a bateauman, had their roots deep in the black dirt of the Connecticut River Valley, back to a time when the Abenaki Indians named it the Long Tidal River - and pronounced it "Gwenitegok."

Yes, native and immigrant boys from towns and farms, from near and far, came carrying bucksaws and axes, and they learned to be woodsmen, bateaumen, teamsters, blacksmiths and cooks and to proudly wear spiked boots. They might have come to the river as boys, but most left as men, "Connecticut Valley River Men" or " River Hogs" as they laughingly called themselves. And they carved their way into the history of a mighty valley carrying primitive cant dogs and pike poles in their hands, forging a trail into history hard followed by lesser men.

I invite you to join me now in the spring of 1898 as the "drive goes down." We will be passing many dangerous spots on the two hundred forty mile trek ahead of us and while some places are still there, others are completely gone with the era, like the dreaded Fifteen Mile Falls which now lie quietly under 100 feet of water. But for us, all the "bad spots" will be there just as they were when the log drivers challenged and confronted them and found a way to beat them one-by-one.

The characters in this book are fictitious, but the never equaled feats accomplished by the New England Log Drivers every spring, and their unprecedented contributions to history, are real and will never be changed. Who could tell the whole story? Who could ever tell the individual accounts of the thousands of men who drove logs and later pulpwood down a thousand brooks and streams? To name or recognize one hero is to leave five hundred men unnamed or unrecognized. Each log driver had his own story, his own legacy. He was like the nameless rushing streams running off the mountains, called by the river and swept on to the sea. The recorded facts are too scarce and that is why the entire story of New England log driving has not yet been told. And I did not attempt to do it. But in this portrayal, I have strived to bring alive to you the likeness of that historical era and the unforgettable part that the northeastern log driver played in that life and death drama. It is in novelization that indeed, history does come alive. And as a living thing it teaches and enables the reader to feel, and sense the era or the subject in a way that no documentary ever could.

The spring log drive was the merging of mighty trees and mighty men from two great valleys, the Ammonoosuc and the Connecticut. It was a herculean feat achieved by mortal men.

The Connecticut River Log Drive began with the wake up call, "Get up boys. It's Daylight in the Swamp!" and that drive ended as the first arriving logs were locked into the booms at the sawmill at Mount Tom, Massachusetts, two hundred forty miles below First Connecticut Lake, the starting point. The cry there at the end of the drive was one of victory,

"God Damn Them, They Did It!" If you find this

statement coarse or strange, it was to the weary river driver a wonderful sound. It was in fact the praise he had longed to hear for endless weeks and months, from snowflakes to black flies. Come Hell or high water, come big snow or little snow they had ridden the deadly rapids, walked the booms, laughed and scoffed at danger and out-roared the thunder. They had done the impossible. They had beaten the "Old Connecticut!" And, they did it successfully and annually for over half a century.

The characters in this book never actually lived, but the annual Connecticut River Log Drive was real and what was accomplished by the Connecticut River Log Drivers was also real and their spiked boots made a mark in the annals of New England history that remains unequaled.

This story could be the life and times of any of the actual river men who lived and died but not before they conquered the river. The real-life heroes are on the other side now. They have passed on to another time and place, but along their way they made history, their own never-to-be-forgotten history. Many of those unsung heroes couldn't write their own names, but all of them heard and heeded the call. When that "Long" River called a man's name, its voice was loud and as real as the lush forests and the pristine streams and lakes flowing into the Connecticut River. Like the log drivers of old, this magnificent valley was alive, alive and filled with a zest for living matched only by those adventurous river men who moved the mountains to the sea, left their sweat upon the land and walked upon the water.

From letting "Daylight in the Swamp," to "God Damn them, they did it!" They did do it! And may God bless each and every unsung one of them because they are the only breed of mortal men that ever could.

Trudy Ann Parker

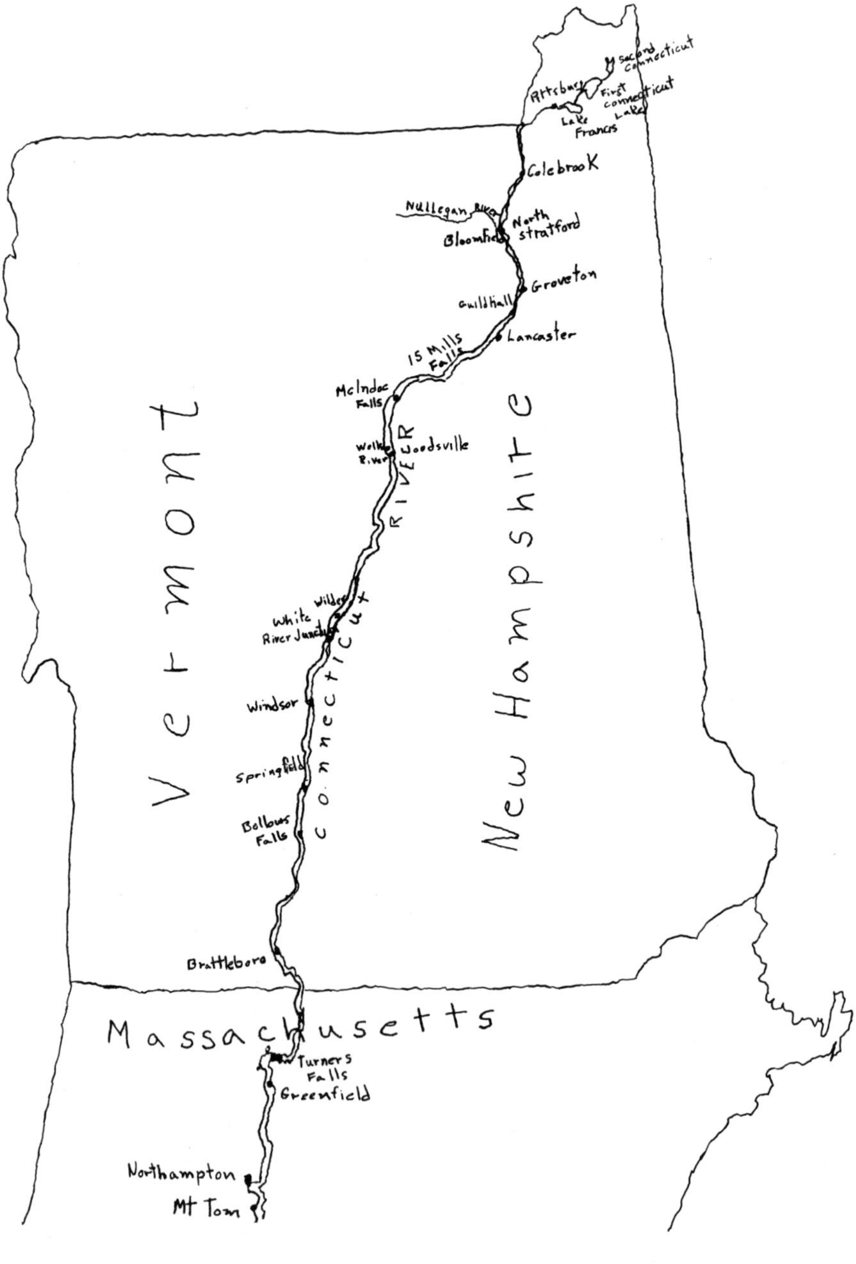
Second Connecticut
Pittsburg
First connecticut Lake
Lake Francis
Colebrook
Nullegan River
North stratford
Bloomfield
Groveton
Guildhall
Lancaster
15 Mills Falls
McIndoe Falls
Wells River
Woodsville
RIVER
Wilder
White River Junction
Windsor
Springfield
Bellows Falls
CONNECTICUT
Vermont
New Hampshire
Brattleboro
Massachusetts
Turners Falls
Greenfield
Northampton
Mt Tom

Chapter One

Indian Stream, N.H. 1898 — It was the last day of December. Numerous lamps flickered and tonight the men conversed little as they sat around the drafty camp this bitter, cold evening. "Jake" should have been back by now. Jake Wolwas was the 19-year-old lumber man who had been promoted to the camp's meat provider. "Beaver" Jones tamped some tobacco into his pipe and was content to listen to the wind which howled and bellowed outside the isolated logging camp. This hovel was made up of a kitchen or cook-shack and combined dining-room. The cook-shack had been constructed first and this was made of unbarked logs with mud or moss stuffed or chinked into the crevices between the logs. The men's bunk-room or the bar-room or ram's pasture had been added on later. The cook-shack had been originally built with a dirt floor but now all the rooms including the little one to the side called the office had floors of poles. Board floors beat pole floors but pole floors beat dirt ones.

The filer had a separate shack and there were two other ones for teamsters. Downstream from the main camp were the horse hovels and there was a five-hole back house with an open front. The only season a man could while away the hours was the same one where the black flies chewed and feasted on his flesh.

The wind kept these back houses well ventilated and that open front allowed the rains to keep them washed. Speaking of the out house, now and again various men got up and pulled on a jacket, touched a match to a lantern and made that lonely trek down the path. Beaver heard the wind moan. It seemed to be dying down some but when the door was opened by the outgoing or incoming loggers, snowflakes flittered around. God, there was times a man was lucky to be constipated, thought Beaver, and this was one.

At that moment, a wry and infrequent smile sneaked across Beaver's ugly, unshaven face. He was remembering when old Charles, who had fought in the Civil War and with anyone else who would ever fight back with him, had gotten drunk and stumbled out to the backhouse, and there he sat the next morning, in sitting position and straight as a deacon, but almost frozen solid. Only the whiskey he had consumed saved him from dying of hypothermia. Old Charles left a fair ring of the hide off his buttocks in the backhouse as a souvenir. It took two men to move him and another to carry his pants and boots, and once they got him out of them, they too were in sitting position.

Built on to the dining-room was the bunk or bar-room, called the ram's pasture. Here the men slept but first, like tonight they sat on their bunks or crudely constructed stools and puffed on their pipes quite frequently while conversation was almost nil.

There was no light out there anywhere now. It was dark, dark as the bowels of hell and the men were all wondering where in hell Jake was? Of course this snow didn't help. It seemed to the men in the camp so far away from bright lights that it had snowed every God Damned day this month. At least that's what Beaver claimed, and he'd been "keepin' track," they all knew.

Beaver puffed his pipe. It was funny, but the day after they buried old Charles was the day that Jake walked into camp, dressed in farm overalls and carrying a rifle. He was a good looking youth, tall and quite dark complected for a Yankee. Jake was 16 and had left home after his drunken stepfather had undertaken to beating up Jake's mother. The long and short of it was that Jake had pounded hell out of his stepfather who had spit out two teeth, but retained enough of them to order Jake off the farm in Maidstone, and so he went.

The men all liked Jake, and they soon learned why he carried a rifle. He could use one that's why, and no logging operation that gave that young man a job ever had a meat shortage. He just wouldn't allow it, like now, he knew there was a storm coming, but out he went anyhow. Now, Jake didn't talk, "No more than a Galdern Injun," said the men, but that was fine because a logging camp needed doers, not talkers.

The silent crew straggled in and out the door, on their way to the privy and others sat in half stages of undress. On clothes lines strung across one end of the room hung woolen pants and shirts and wet stockings, stained and reeking. The air was sharp with the smell of the woodfire and the aroma of tobacco from many clay pipes and that earthy, sweet smell penetrated every nook and cranny of the low roofed rooms. The men smelled of wood ashes, and they were glad they were inside on a night like this.

The moon swung higher now over the mountains. The storm had subsided but still something was missing. Where were the wolves that had made three raids on the storage room? Why were they not out there tonight, slinking about and howling that they were hungry and were coming in close looking for something to eat? They should be out there trumpeting unless they were not hungry because they had eaten already. Beaver winced. Jake would not give up his daily meat kill without a fight and no one had heard any shots. Now nobody could hear any wolves prowling about either. It was most unnerving.

Just then the door swung open and in came the missing man. His normally dark eyelashes were white with frozen sleet. "I got turned around in the squall," he said. Across his shoulder he had a young buck but one of the hind quarters was torn and jagged. "Help me with this. Until a few minutes ago, I had a Goddamned wolf hanging onto this buck's leg," said Jake and his voice was sharp.

As men sprang forward to help unload the carcass, a voice asked, "Why didn't you shoot the bastard?" Jake looked at the men. "I'd used the last bullet. I hit him with the stock but him and a big female made off with part of the leg."

Feet shuffled from every direction and hands reached from everywhere to help Jake. Two of the men took the deer and started for the woodshed where it would be dressed by the light of lamp and lantern, while the camp carpenter and also butcher slapped his cold hands together to keep his fingers from getting stiff and untrue with the skinning knife.

Peatoo brought a large helping of left-over dried beef and cream, which the younger man ate greedily. Beaver sat in the shadows and puffed and finally asked, "How'd you ever get yourself home after dark?" "It was the train whistle that got me home. The wind had been blowing a gale and then it died down some, and I heard just those Christly wolves is all. But, then I heard it just faint, and I recognized that sound. It was the wail of the whistle of the old Maine Central making her run up the grade heading through the valley."

Heads nodded and men chuckled. For decades North country residents remembered the shrill scream of that grand old train, as it rolled into and out of the depot at North Stratford. This was the junction with the Maine Central. The Grand Trunk was a mighty mogul that chugged and moaned in sunshine or raindrops up that rugged and once considered impassible landscape, the Nulhegan Valley which was the same one Jake and other locals called the 'Hegan Woods'

"By Damn, I know where I am," continued young Jake, "I'm headed east and I'm not far from camp, I said and I started out but them Goddamned wolves got their noses into things. There was eight, but there is just two now, and them two Christly bastards got away with the leg I planned for supper." That was a whole lot of talking for Jake and the men laughed, because about the camp it was not often that they heard him speak, let alone swear.

His supper eaten and partly digested, a tired Jake pulled off his wet boots, he was exhausted. This day had taken its toll on him. Once more before he shut his eyes he thought he heard the goodbye whistle of the train. Right at this moment she was bellowing and pounding up that long grade and when she once started down, she was headed for the waters of the St. Lawrence, and nothing this side of Hell could stop her. Jake knew that the very trail that she followed, right where those steel ribbons lay, was once roamed by countless Indians. Those Native People walked softly and slipped silently into posterity unnoticed by many historians who ignored their presence, seldom hearing or heeding what little they did say.

Once more the lonely whistle could faintly be detected by keen ears. It was a haunting and ghostly sound and somehow it always reminded Jake of his Indian ancestors. All of Nulhegan country was the home and hunting grounds of one side of his family. His name was not Jake, it was Philip and he was the fourth Philip belonging to the family of the one the whites called King Philip, Chief of the Cowasucks. King Philip made a deed in 1796 in the season of Niben (early summer) June during the Nokkahigas (Moon of Hoeing) the old ones said. He had already seen nearly eighty winters and some said it was because of his advanced age and fear of not being able to any longer provide for himself, that made him grant to what later became the Eastman Co., all the lands north of the forty-fourth Parallel including granted lands as far north as Canaan, Vermont, and Stewartstown, New Hampshire. This was one tremendous tract of land stretching from Fort Wentworth on the bluffs high above the junction of the Connecticut River and the Upper Ammonoosuc River and continuing all the way to Lake Memphremagog and on to the Canadian Highlands.

Many in his own family said that King Philip had no right to make deeds like a White man. They said he was self-serving and not thinking of his people. They said Philip was wrong to sell the land because one man does not own all the land. Others believed Philip did think of his people and of their ancestral hunting rights, but they also felt he trusted too much the White people he sold the land to. They were the ones who did not honor King Philip's wishes, the hunting ground requests that would have given his people and all Native peoples hunting and fishing rights to this land forever. These stipulations Philip believed were a part of his deed, the one called Philip's Grant. This Chief was unable to write for himself his requests and so they never appeared as part of this massive land transfer. And so that controversial man, that fearless warrior who could not be beaten by his enemies received the defeat of his life at the hands of his "trusted" White friends.

This was much talked about when Jake was a child by his great grandmother. She was the one who carried King Philip's blood and she stood firm he had been betrayed by men he trusted. He was not the betrayer, he was the betrayed.

Jake decided that Philip was a man who looked out for Philip and to Hell with his people and their hunting rights. He reasoned that old Philip didn't think like an Indian. He thought like the people who had called him a "King."

Those who claimed Philip reserved hunting and fishing rights for himself and all his people forever also admitted that he had a strong thirst for rum.

Jake did not know how he felt about this. He guessed he was in sympathy with Philip's people and not the man he was named for. But, if King Philip had not done this, would Jake have a job? He wondered about this. To have lost the hunting and fishing rights was a bad thing, but to have a job and be able to better your life was a good thing, so was Philip right or wrong? Jake twisted and tossed trying to get comfortable. Just before he closed his eyes, he caught the sound of the last faint whistle of the train as the Grand Trunk pitched over the top of the valley en route to Montreal, Canada straight ahead and going down.

The men who were skinning the deer came in to say that those wolves had done a job on that buck's hind quarter. They allowed Jake had done all right to get out of that fracas without loosing his life. Besides they still had three legs more than the wolves did to chew on.

Bill slapped his cold hands together and said, "It's building up to storm again. Its a good thing that boy heard that train whistle." The men agreed.

Old Beaver couldn't remember many a night when he hadn't heard the eerie warning of the Grand Trunk, but he hadn't been listening for it tonight, he had been listening for the boy. It didn't matter if the camp hadn't heard that haunting wail tonight, just as long as young Jake heard it and got his bearings.

It was said that the Grand Trunk's wail could wake the dead from their long sleep, and yet it was sweet music to the living, for it told all the isolated inhabitants of the river towns of Brunswick, Bloomfield, Lemington, Colebrook and Canaan that with a full head of steam that old engine was grumbling up or down the valley. This railroad line, between Portland, Maine and Montreal, Canada brought woodsmen to the logging camps and tourists to the sea. And tonight, it had brought one young man home.

Chapter Two

"Get up boys, it's daylight in the swamp!" That was the early morning wake up call that greeted the woodsmen daily. This was New Year's Day and somewhere ladies and gentlemen slept late catching up on lost sleep, after the gala balls that welcomed the New Year. That was not so here on the banks of this frozen stream.

It was bitter cold outside and the warm quilts felt good, but it wasn't Sunday, and therefore no one employed here would be sleeping late.

The cook and the cookee had been up since four o'clock in the morning and there were good smells radiating around the dining room. Coffee was the strongest of these pleasant scents followed by bacon frying crisply in a great stove-top skillet. Later, when the pan cooled down, eggs would be fried in the bacon grease with the cook making sure that each yolk had bacon grease spooned over it. There were stacks of steaming pancakes, as big as dinner plates with plenty of maple syrup to pour over them. There was a choice of corn or date bran muffins and a separate pan of scrambled eggs for those who didn't like them fried and there was hash brown potatoes with onion, browning to perfection and waiting for the hungry men to take their places along those wooden benches.

About three quarters of the men in this Indian Stream camp drank tea rather than coffee, but even those who didn't like the taste of it accepted or endured the smell of it. This smell in the morning, like home-made bread toasting on a stove-top grate was a welcome one.

The window panes were frosted and the day out there would be just like any other, but come Sunday, beautiful Sunday all the men could rest.

The men ate in silence. There was little talk. There was no time for talking when there was daylight to be let into the swamp. And with full stomachs and aching muscles the woodsmen finished this first meal of this new day and put on warm coats and mittens and started for the great spruce and fir forest that would be many trees short by nightfall.

Once out of doors the frigid air of this first day of January made sharp contact with this hearty group of men carrying axes, saws, and cant hooks. They started for the woods then, one behind the other. On a day like this, it was like everything happened in slow motion.

The axes that felled the trees had a harder time to get going. All day the timber would fall on this mantle of snow covered earth, and there would be almost no sound as the great, tall trees hit the ground. A spray of mist and snow would shower the air but the trees seemed to fall to the frozen earth in a muffled silence. It was like in passing, they had no voice. Their death song should have been louder.

Frequently the men would stop and slap their gloved and mittened hands together, to overcome the numbness of their tingling fingers.

The sun came up slowly and it seemed even that fireball was slowed down by the bitter cold.
Already the teamsters were sighted at the edge of the clearing ready to haul the great logs to the landing on drays pulled by two or four horses. Steam could be seen coming out of the mouths of the great work horses. They chomped at the bit and snorted and didn't seem to be any more comfortable than the other creatures working there — the men.

The sun moved higher in the sky. It was clear and cold but there was no wind. Wind was what was dangerous to the woodsman. When he notched a tree, if the wind changed, the tree could fall the wrong way. Then there was the ever-present danger of a limb falling and hitting a man. It was known that no man could harness the wind and that only the wind, not snow or cold could prevent the woodsman from doing what he was paid to do, "letting daylight in the swamp."

This day had begun early when only dim rays of sunlight could be seen and felt, but now the sun was brighter and it bathed the snow-covered forest floor with first light. The stiffness was already working its way out of the men's hands. They were limbering up, and as they toiled with their axes and buck saws, they like this day itself, were warming up already.

The thunk, thunk, thunk, of an ax against a great spruce and the whine of the two-man crosscut saw echoed across this frozen tundra and blended in with the snort and blow of the twitch-horses as they made their way toward the landing below. The runners on the drays that hauled the logs had once been logs themselves and now they squeaked and screeched across this man made road of hard packed snow.

There would be no break until the sun was high overhead and when that time was finally here, Jerry the Cookee would be arriving carrying baskets of food and the lumberjacks would stop long enough to eat sandwiches of corned beef and in hard times even made of bacon or boiled salt pork, cold baked soldier beans and slices of raw onions. There was frequently huge molasses cookies or tasty sugar cookies and always for the men there was tea. It was preferred strong and black and it sustained the woodsmen.

The crew would eat with great relish and then take time for a smoke while lunch was digested and by the time the sun, that boss who never rested, was no longer directly overhead the men knew that it was time to go back to work. From sun up to sun down went the workday, but as long as there was good food to eat and strong tea to wash it down with, what man needed more except to sit at night reflecting and smoking his pipe as the stillness of the darkened forest wrapped herself around him?

All day the trees fell and the crews that limbed them crawled over the logs as easily as that squirrel whose residence had been destroyed when the tree came crashing down in a cold and deadly silence.

This day was unremarkable until mid afternoon when great bellowing and cursing could be heard coming from the landing where the logs were being twitched.

A new man was helping unload the dray and a belly-girt strap broke just as the spirited team arrived at the landing. The spooked horse had jumped against the harness and its mate which was equally frightened had also jumped. At that exact moment a great tree top rolled free from on top of the load and it shifted or (hauled) and other logs began to shift and roll while men at the landing who had been laboring with cant hooks and others bent over their cant dogs yelled and took off in all directions, all but one Nova Scotia Scotsman, known to the crew as "Mack." He was in an awkward position, off balance from the start he could not flee the landing and his leg was crushed between two logs.

His scream echoed through the forest, piercing the stillness, and then there was only the sound of silence. The teamster pulled on the reins to quiet the bolting team and other men came running to use their peaveys and cant dogs to free Mack's badly mangled leg. It was there, hanging at a curious angle with blood spurting everywhere. This was a bad one.

When the horses had been quieted and the logs at the landing had stopped rolling then muscles swelled as the burly woodsmen strained and grunted to roll the confining logs and free their friend's crushed limb.

"It's almost out Mack, we got her, we got her," big Frank Lansing encouraged his friend.

As the crew worked the injured Mack was in a semi conscious state, draped where he had fallen suspended by the leg and by the two great logs that held his broken limb like a vise or a bear trap. The men came swarming around like bees now, watching the logs that had rolled up behind the one that had pinned the leg. They must go easy here, or more than one log would roll and more than one leg would be crushed.

They pried with their cant dogs and rolled and held fast to the next log threatening to roll down from the loaded dray and finally they succeeded in freeing the broken leg. Some of the men threw their jackets on the snow and they laid Mack on top of them. He was going in and out of consciousness and asking if they had got his leg free. He couldn't feel it.

Someone brought a great heavy felt and this was placed over the injured man and an ancient horse blanket served as a stretcher that carried him back to camp. He was bleeding profusely with part of his shinbone protruding through the skin.

"God Damned compound fracture," muttered Beaver as he watched the figures moving off toward the camp with Mack. He knew that this was the worst possible kind of injury, a wound that was hard to clean, and even harder to heal, if it ever did. He shook his head and wished that he did not know what a potential for gangrene there was here, but he did, and that was that and what would be, would be.

The sun was heading for a setting position and the men had returned to work but the unfortunate accident had taken the zest for working out of most of the woodsmen and their axes hit lighter blows and not so often and the whine and creaking of the saws could be heard but not so steady now. The enthusiasm was gone and it was almost time to call it a day. Soon the crew would be going home to find out how Mack was holding out.

Everything that could be done for the man would be done at the camp, but there was no doctor here and only untrained men to attend to the injury. They were not medical professionals and they would be trying their hardest to salvage a man's leg, based on what they had seen in life and what they had to work with.

They would give him brandy or whiskey to drink to kill

the pain, and pour raw alcohol over the torn and crushed flesh, but that did not heal broken bones. If Mack needed an amputation, he would be undoubtedly sent by a camp sled to a doctor in nearby Colebrook, New Hampshire.

The men returned to learn that he did not, the untrained but sterilized hands of the cook and one of the saw filers cleaned the crushed bones, dug out all the bone fragments that they could and finally got the damned bleeding stopped. And here is where the expertise of each man they had came into play. Young Jake was a dead shot, but Bill the camp's butcher was the man that was good with a knife, a skinning knife or even a small buck horn hunting knife, like he used to do the precision cutting and scraping as he cleaned away the tiny chips of fragmented bone and clothing that had gotten ground into the wound. If you haven't got a surgeon, get a meat cutter and they did.

They swabbed the injured leg with raw alcohol and would dress it with bag-balm and clean bandages which they changed daily, and they would all hope that red streaks would not proceed the green and putrefying flesh that became blood poison.

Mack had gotten to North Stratford via the Grand Trunk railway and he had walked all the way into this camp. He had walked in, but it seemed that he sure as Hell would be limping out.

Would he be able to swap his ax for spiked boots and ride those big logs to Mount Tom Sawmill in Massachusetts, later this summer? That was the question on every man's mind. Would he be able to take the drive down? You're damned right, he would, if he didn't die of the gangrene, and by Jesus he came here to drive logs, not to die! He told the men that from his cot where he lay sweating profusely and sipping hot brandy.

This was a good sign, a damned good sign thought young Jake and old Beaver and all the rest of the men. This was more than a woodsman — this was a river man.

"If I can't ride them Christly logs with two good legs,

then I'll whittle me a sturdy cane with a pike pole on the end of it and ride em with one!" And he did just that, that Nova Scotia Scotsman, on the spring drive of 1899 and for two more drives after that before he succumbed to married life, traded his spiked boots for a hay fork and became a low profiled, highly respected north country farmer.

Chapter Three

Isabel Lauren MacCutcheon was a petite woman, slight of build and two full inches short of the five foot mark. But this had never been a problem nor a detriment to Sable, because what she lacked in size she made up for in spirit, and this woman nicknamed "Sable Lore" was indeed spirited. She had been given brains and also a tongue that could cut to the bone, when she turned it on someone who had angered her. Her eyes were a strange shade of blue, often called hazel-blue, but they were not. They were green, green with blue flecks, the very color of the stormy sea from which her ancestors came. She had a winning smile that at times won out over her temper, causing her to laugh rather than rage. Her doting step father to tease her, often said that she was "very even tempered, mad all the time."

Her hair was on the auburn side red-streaked blond and her impish face was very pleasant to look upon with high cheekbones and delicately chiseled features. She was a woman of high color, not pale like so many a lass of Celtic roots. Sable had a deeper skin tone, and because she was so healthy there was always a natural rose tint on her cheeks.

Sable's childish pout and sharp glance gave fair hint of the anger that could rise up within her. She had a way of looking haughty and disdainful when something displeased her, and people said when she was angry she looked more like a panther than a woman, her wrath so readily seen.

Sable was delicate in stature, and small for her age, but that had never stopped her from being the bully of the school she attended. During childhood days when Sable, fists flailing, got into a brawl, she had a way of throwing her adversaries off balance, knocking them to the ground and then springing through the air like a wildcat. Once she had her opponent down, she had been known to crawl right up on top of them and pommel them with her small fists. And in at least two cases she had to be dragged away by a shocked school master, because she had stopped pounding and gone to choking the boys, whom in each case had provoked the attack by calling Sable names. This violet behavior and explosive temper was what decided her parents to send sable to a private school for girls. She had very few friends there and no one to fight with, so she endured. Sable's parents were sure that she would return to them a refined lady. She returned to them as Sable. They despaired.

Sable had MacCutcheon blood in her veins, and it was from her real-life father, Roscoe "Cud" MacCutcheon, that she had inherited this dark side. And this dark side had made her gentile, adoptive parents extremely sad. You see, they could not change this willful child, later the young girl or lastly the obstinate woman. No one could.

Sable's banker father, Robert Bigelow and his wife Mary would never have had a child to care for had Robert not stopped into a bar in Montreal, on his return from a business trip. As it happened on that stormy October night, old "Cud" MacCutcheon, the captain of a small merchant ship had arrived back in Montreal. The wind was howling and snowflakes were swirling around his curly, red hair. He stepped in out of the cold and out of the darkness into the light of the barroom. Cud squinted, and looked over the group.

Robert Bigelow, that man who would become Sable's father, looked up from the table where he sat drinking Scotch alone at the back of the room, and his first impression of Cud MacCutcheon was the only one he or anyone else would ever see. Cud, like his daughter, never changed. What you saw was what you got. And what Bigelow saw was a powerfully built little Scotsman, who looked more like a pirate than he did a man. This pirate did not have a blue beard, but instead a red one because his bushy beard was even a deeper shade of auburn than was his hair.

Cud had been hurt in an accident at sea when he was a young man and his left leg was badly mangled. It should have been amputated below the knee, but Cud threatened to kill the ship's doctor and in his own words, "any other son-of-a-bitch that let that butcher saw his leg off!" There was one notable thing about old Cud though, and that was that he did not lie. What MacCutcheon said, is what MacCutcheon did, and for this reason the useless leg was left intact.

The feeling of his shipmates, older and younger, was that a limb as mutilated as that one would never heal and Cud would mostly likely die of infection or gangrene, but Cud knew better than that, and he healed in spite of all their doubts. Thus, with the help of a sturdy cudgel that served as a walking stick and a war club in battles, he managed to lead as normal a life as any other drunken sailor, except that this one rose to captain of his own small ship.

Cud could lick any two men without his trusty cudgel, but with his 'cud', he had been known to pile them up three and four high on the barroom floor. And brawling anywhere was what he liked to do best. It was from his well documented proficiency with the cudgel that Roscoe Angus MacCutcheon got his well-deserved nickname which followed him through the long years of his colorful life.

Captain MacCutcheon had liquor on his breath when he walked up to the bar and ordered more. True to form, he carried his thick stick but tonight he cradled against his left breast a baby girl wrapped in many blankets. This infant child was his daughter. She was four days old.

When Cud had brought his ship into Montreal, he was greeted by news of a healthy girl child and a dead wife. He was also told by the port authority that because of his shady dealings that he was to leave this harbor and never bring his beloved boat "the Glasgow Belle" into this port again.

In the dim light of the barroom, Cud's hair was seen for its true color, which was a dark, strawberry-blond and his disposition came out too, showing his true color also.

With glossy eyes, Cud gazed around at the well dressed gamblers and businessmen, with their kid gloves and top hats resting near. They sat quietly and watched Cud, and he stared back at them. Cud thought to himself, Scotch whiskey, it blends all modes of men, rich and poor, purists and pirates, putting them all on even ground. A smile curled Cud's lips when he discerned, Men were Men, the only difference being that some could drink more than others, but all men appreciated fine Scotch . . . the gentleman and the ruffian.

Cud sipped away, a shot glass at a time, until the Scotch in the bottle shrank about one half, and then he came fully erect and addressed his drinking comrades,

"Ladies, Gentlemen and others, and I'm sure there are some of each here. Now would ye notice this fine girl child that I have in these blankets." Cud eyed them all and then went on, "Now if this child of mine was a boy bain, I'd take him to sea with me, but being a girl, she would only get in the way and I could not care for her proper like."

The barroom had lost its noise as a strange hush fell over it, and all eyes were glued on old Cud. The man at the piano had stopped playing and one of the barmaids came over and peeked into the blankets and said, "Oh, such a cute little mite," followed by the question, "Sure and you have no relatives that would take this babe, not even one?"

"Not even one," replied MacCutcheon in a pitiful voice, then he added,

"Not a relative to me name, nor a friend either, dear lady."

That doesn't surprise me, you old devil, thought banker Bigelow, from where he sat quietly observing a man trying to give away his child.

With piercing eyes that never wavered, Cud studied

them all. Then without further ado he raised the bundle in his arms, high in the air over his shaggy head and seemed about to toss the baby through the air. Gasps were heard, and now that he had their attention MacCutcheon bellowed,

"Last chance to get a fine babe. I'll swap her with any fine gentlemen or lady here for a bottle of Scotch." They were thinking now. Cud grinned.

Robert Bigelow stood suddenly and said, "I'll take her." He looked at the bar keep and said,

"Get the man his Scotch." Bigelow walked over and paid for the bottle and Cud MacCutcheon reached greedily for that bottle with one hand and handed over his own flesh and blood to a stranger with the other.

MacCutcheon's eyes were dark and stormy now, and they fixed on Bigelow and he said clearly,

"Her name, good man, is Isabel Lauren MacCutcheon, and that she will be told someday when the time is right."

Robert Bigelow nodded and replied, "That she will, and even more." He had replaced his top hat and now with his expensive kid gloves, he reached for this infant, tucked her under his arm and headed for the door. Mary would be happy he reasoned, for they had frequently spoken of adoption but what this banker wondered was, when would the time be right? Would there ever be a time that was right to tell this child, or any other, that at four days old she was swapped at bar in Montreal for a bottle of Scotch?

Now years later and much water washed over the dam, Sable knew. She had been told the whole story at the ripe old age of thirteen. She had received the news with little or no emotion showing.

She often had felt the Bigelows were not here true parents. How could they be? They were nothing like her and she was nothing like them. They were cultured, and gentile and everything she had no patience with. Her real name was not Bigelow, it was MacCutcheon. She was Isabel Lauren MacCutcheon and she felt more like that name already.

Sable Lore watched the pain in her step-parents' eyes

as they told her this truth, and then her sea green eyes seemed to look almost through them. She nodded in acceptance and a demure smile adhered itself to the corners of Sable's lips and she remarked,

"Aye, and it must have been a rare and fine blend of Scotch indeed, to have been worth so much as I."

Chapter Four

Upper Coos Meadows - It had been a bleak April day and like no other Jake had ever seen. He knew that he would remember it. This would be his second year "going down with the drive" to the sawmill at Mt. Tom, Massachusetts and he had seen men drown en route, but not one who had visited with him just hours before. Jake had enjoyed the conversation too, because not being much of a talker a man learned a lot. The young man was Irish with strawberry hair and merry, blue eyes that twinkled when he told a story, and he had told a good many last night. His name was Patrick something or other, but everyone called him "Red." Ironically, he was a year younger than Jake and already had two drives under his belt and he expected to complete his third.

How he had made the men howl with laughter, when in the flickering light of the little fire in front of the white tent, he told them about his first experience at a house of ill repute Black Widow Mudget's Whore House down in Holyoke, Massachusetts.

It was Red's first drive and the older and more experienced men had set the whole thing up. "She ran a good one," Red laughed and went on "No one ever had to worry about getting the creepin' crud at the black widow's place and I didn't get it either."

Jake was forced to grin, when he thought of how smug his friend had looked as he made that statement. The story went that the black widow's business establishment was located at the very end of one of the many "Red Light Districts." Concerning those infamous streets, they used to claim that the farther you went the tougher they got, and widow Mudget lived in the last house.

None of the widow's young harlots would be right enough for Red, the older men thought, so they paid for the services of the black widow herself. They told Red everything he needed to know about the widow Mudget except one thing, that she was a black woman, really black with long, spider-like legs that let no man get away, not even Red.

Red claimed when he got to the dimly lit room that there she sat. She smelled of perfume, strong perfume and wore some kind of a gown with feathers on it. He swore that she blended right in with the dark and it took him a few seconds to get his eyes adjusted to her as she was so black that she shined. "She was blackern' a boot and hottern' a cannon!" Red exclaimed.

She motioned the young man to come nearer and he did. He had bathed, but she could still smell the woods on him. The widow, a huge woman with thick, red lips smiled pleasantly and tapped on the bed. She told him anything he wanted to do, they would, and anything he didn't want to do, they wouldn't. Red thought this was a pretty good deal even if he hadn't counted on one so old or so feathered. "She was as old as my mother," Red said and went on, "and in the closer lamplight she wasn't too bad lookin' and after she fixed me a drink, she was better looking. By the third drink she had three tits and was beautiful!" Again how the men roared with laughter.

To make a long story short he made very good use of his time. Red claimed that she was partial to lumberjacks anyhow because the late Mr. Mudget had been one. She commented that Red was a real gentlemen for a river driver. She also said, "Well Red, you are a rare river man. I usually have to ask them to take off their spiked boots before we begin." And this struck Red funny and she and he had a great laugh over it and then according to Red, they got right down to business.

Red had taken notice of a girl with auburn hair he met on the stairway going up to the widow's room. The stairs were narrow and she smelled like lilacs. She wasn't too big and she wasn't too small, and she giggled as she brushed up against him. He forgot he was nervous and all he could remember was that he liked the way her nose turned up. He vowed the next time he went there he was having a go at that little spider and not the old black widow!

Most of the "Mudget Girls," were mulattos but the one Red met on the stairs wasn't. Her name was Molly O'Grady and she was as Irish as he was. While buttoning up his trousers, he asked the black widow about her and she teased him, "You come to pleasure yourself with her — you will need your spiked boots!"

Speaking of spiked boots, Jake had slept with his on. He had gotten them soaked and if he took them off, he would probably not be able to get them on again. He wriggled his toes, and they were still wet. He had rubbed his boots and the insides of his socks with mutton tallow and they were some protected, but when you are in and out of icy water to your waist several times a day, nothing protects your boots. He went to the bushes for his evening toilet and from there made his way to the little sand bar up ahead. Here he bent and began dipping his hands in the water and then scrubbing them with sand. This was the way his mother's side of the family had cleaned themselves for years. It wasn't that he had anything against soap, it was just that living the outdoor life that he did, soap wasn't always around, but sand was.

Old Beaver passed Jake and grunted at him. He had come out of his tent that morning, aching and in a cantankerous mood. He had blown his nose with his thumb and forefinger and then flung the secretion to the ground and wiped his nose on his sleeve. As a matter of fact, there were nose tracks all over that damned, dingy sleeve. The men often joked if filthy was the fashion, then Beaver was in style!

Jake remembered now how comical Beaver had looked when next he had given a spry hop and thrown his right leg up in the air, and passed gas so loudly that he seemed to be giving a twenty-one- gun salute to the cook for the excellent beans he had cooked for supper last night. Beaver hopped around then and began breaking wind loudly and repeatedly for several seconds.

Good morning grumbles and chuckles came from his cohorts around him.

"Let her Rip, Boy!" someone was heard to say and from another tent a new voice said,

"By Damned Beaver! Any fart that loud had ought to be lowered with a rope!"

Beaver who was still hopping around and fanning the air with his old slouch hat informed them all that he was "holdin' back some."

This brought all the men in hearing distance into a low rumble of laughter but as for Jake he believed that old Beaver was truly "holdin' back some" because if Beaver ever did cut loose, and fart to his true capabilities, the whole damned shooting match of them would probably land in Turkey!

It was a known fact that passing gas or (breaking wind) was as much a part of log driving as was grunting, bellowing, straining muscles and sweating. These were the sounds of life as the drive went down, and it was a good life, one that not a man of them would swap with any other for almost any price. There was a freedom to this work, a man could swear when he wanted to swear, burp when he felt like it and fart to his heart's content, after all they called themselves "river hogs."

Jake once heard a bateau man say that there was no peace on the river, only the way he said it was, "The men bellow, curse and fart by day, and them Goddamned frogs, peepers and loons bitch about it all night." And this was the gospel truth. Another gospel truth, is that life goes on, and sometimes it takes the likes of old Beaver to make others realize this. Dead men could not do what they had to do, which was take the drive down.

Now it was nearly dusk and the low rumble of thunder could be heard. Jake came back to the present surroundings, a cold meadow, fast running river water, with chunks of unmelted ice and snow floating in it. Well, he guessed there was no need dwelling on what had been nor worrying about what would be. Red was a reputed good swimmer, but on a spring drive, a good swimmer could find a watery grave just as quickly as a poor one if a man was distracted and not alert at all times.

Jake would not allow himself to think that Red had been careless but he did wonder if his mind had been somewhere else. It couldn't be anywhere else on a log drive.

The log Red had been standing on twisted and then rolled like hitting something beneath it. The jolt threw him up and he struggled to regain his balance but consequently he fell into that water like a brick. He fell at a spot where the undercurrent was strong, and the water black and deep. He bobbed up once, and the men tried to get him but he went under and never surfaced.

They fished his body out of the water downstream. With rough hands that were as gentle as a woman's, they tenderly removed his spiked boots. And while tears rolled down many faces they rolled their deceased friend in a heavy felt. Red's body was then loaded into the back of a supply wagon, and it was sent on ahead to North Stratford for shipment home on the old Grand Trunk. They hung his spiked boots on a tree and as the clatter of the horses' hooves faded away a great hushed silence fell over the noisy meadow once more.

Some of the men were wiping their watery eyes and noses with dirty bandanas but as for Jake he only stared after the wagon and made a promise to his friend, "I'll do this one for you Red."

That Irish lad, so young and full of life was going home, for him the drive was done, but for Jake and the rest of the men in scarlet wool shirts, the drive was straight ahead and going down.

Three nights after the tragedy found the men camped directly on the meadow across from the Town of North Stratford, N.H. This village was such a welcome sight, it was a town alive and these men needed something to take their minds off death. North Stratford was a railroad junction and wagons ran day and night bringing supplies to farmers and logging camps. It had plenty of just what these men needed, women and whiskey. And, if there was anything could take a man's mind off death it was weak women and strong whiskey.

The supply wagon called the "wangan" had brought them a good supper, but it could not bring back the friend they missed.

This fatality so early in the drive had shocked everyone. Good Christ, if they couldn't handle the Coos Meadows how in Hell would they handle the Fifteen Mile Falls, when they got to Fitzdale or Gilman, (Lunenburg) Vermont?

The temperatures dropped sharply and the welcoming lights of North Stratford twinkled. Tonight many a man would sneak over that side of the river, and get what he needed. After all, the untimely demise of Red had shown clearly how quickly life can be snatched away.

Jake didn't usually go, but he decided that tonight, he just might. He didn't need a woman, and he really didn't like the taste of whiskey, but tonight he might go and purchase some because he could, and it would make him feel alive. You see, Jake needed to feel life this night, when he had felt death so close to his heels all day.

Chapter Five

North Stratford Depot — She stepped briskly off the the car behind the "old Grand Trunk" steam locomotive as it chugged and squeaked to a clanking stop. The smiling conductor handed Sable Lore MacCutcheon Glenhill Dupree the one good sized valise that she was traveling with. This contained all that the little woman owned. It would have been a sad situation if said valise along with a few choice articles of clothing hadn't also held a metal box filled with more money than most people would ever see, including higher-ups on the drive.

It was growing dusk and having been a dark day, good light was fading fast. Sunset, thought Sable, this was a Hell of a time to be starting a new life. It should have been in the morning. Oh well, that would come tomorrow, if she could find a place in this one-horse town to sleep tonight.

Everywhere Sable looked, she saw lumber jacks strolling about. Some of them passed quite near to her on the busy depot platform. Most of them were as big a trees, thought Sable and the smell of pitch and of the deep woods itself radiating from their clothing was not unpleasant to her. Yes, Oh yes, this refined young woman who looked like a lady had some very earthy qualities.

They were a masculine looking bunch and rough cut

many of them, but when a lady passed, they removed their hats and treated her with respect.

Sable liked this quality in a man. A real gentlemen, with good manners was not necessarily the mark of money, but the mark of good breeding, money or not.

She smothered a mischievous if not devious smile, and she chuckled silently to herself because she knew that she saw a quality in men that most gentile women, did not — the fools.

It was Sable's hope that these primal specimens of the deep woods were not "lousy as pet coons," as she had heard them described often.

She passed short ones, tall ones, clean looking ones and a couple that would have given a hog a hard one. She knew that this was getting on toward mid April and the drive would be going down. She also knew that it was said of these river men that they were most generally in the rutting stage when they hit town, after long, lonely months in logging camps. Months of solitude with no female comfort, stashed away in the swamps and dark spruce forests would do that to a man.

Was Sable Dupree afraid of these woolen clad tramping giants as they passed close to her? Hell no, a well aimed hat pin would stop any man, but a God Damned louse (that could jump six feet) was hard to stop.

Sable hurried on her way past the burly log drivers.

North Stratford was a virtual thriving little metropolis. Its lights were such a welcome sight after a long and tedious ride, in a passenger car on the Grand Trunk line. She had heard that they had just followed an old Indian route and if that was so, the old Indians were drunk. She had been all the way across the sea to England, and had never seen so much woods as the trip from Montreal to North Stratford, through the Nulhegan Woods.

The acrid smell of coal cinders was strong in the air, but something else was in the air and that was life. Life sounds were all around Sable and she needed to breathe these, after all she had seen so much of death, burying two wealthy and faithful husbands already.

It was early spring in the northern reaches of the Connecticut River Valley and the annual log drive wasn't starting, it had already started and the van, or front of the drive was already camped across the river on the Upper Coos Meadows, overlooking North Stratford.

Sable did not realize that thousands of logs had already been moved, sluiced (catapulted) by man made water, creating a force to drive or float the logs into narrow sluiceways. Some had been towed across still waters (lakes and ponds) where a log would not float like on a river current.

She did not realize that two deadly, dangerous places had been passed already to get to these meadows on the Upper Coos and that was Pittsburg Falls and Perry Falls. They each had crippled and killed and would again.

Sable did not know all this and if she had, she would not have cared. She just happened to be in a log driving town at the wrong time. No, it was the right time, the precise time when the water was high and the pressure was just right. Now the logs lying in great piles on the landings by the sides of streams and rivers had been released and driven down stream where they had already encountered large rocks and boulders in midstream. The men on the drive would be encountering these obstacles all the way to Mount Tom, Massachusetts. The logs would be driven or actually ridden, through narrow gorges and ledges called "squeeze holes," and this was a necessary part of taking the drive down.

It was always with a great thunderous roar that thousands of cords of wood, cut during the fall and winter, were on their way to market. Market, now there was a word that money-hungry Sable could have understood. But she could not see money and the market was too far away. All that she could see was men, scurrying everywhere like mice. And that brave river driver so respected and admired by the masses was wasting his time showing off for Sable. That doer of the most dangerous work, doing his ballet on the backs of floating logs in white and foaming water and directing, forcing, driving those bullheaded, slippery logs down stream, was nothing short of magnificent, but she would have thought it stupid, work for a squirrel, not a man.

There came a sudden burst of steam and the whistle screamed and Sable jumped and nearly dropped her valise. Christ, that thing needed a muzzle! She regained her footing and made her way through the milling crowd.

The entire valley seemed to be waking up from its long winter's sleep and Sable was quite sure that most of the male population was here, tramping around this little north country town. There was such anticipation in the air as a restless people had waited impatiently for the last ice to melt on the partially frozen lakes and streams.

Spring run off and high water, now it was that the selected timber could be taken down river out of this valley. Now it was that the great spruce and fir forests of the north country would comprise a river of wood as the logs floated and rolled along guided by that elite breed that Sable dismissed so lightly.

Making her way along the crowded street, Sable had already begun to realize that this log drive bespoke of money, money for the merchants all along the route, the entire length of the Connecticut Valley. Ah, the merchants must be grinning like Cheshire cats, she reasoned. She would.

Sable's first plan of business (and she always had one), was to find a place to stay the night.
A cold, spring rain had begun, but this did not depress her, after all, very few things in life ever did for very long.

She had not decided to stay here, but this did seem a likely spot to find some struggling young business and turn it around, into a profitable one. Money was no good if it didn't make more money she believed, and as she thought of this her brows knitted together.

The rain was starting to wet her hair, but she didn't care, it was natural curly anyway and besides that she was thinking. Opportunity seemed to be knocking, but she must think if she wanted a struggling business or to start up something new.

Now what was the most profitable and sure business in a town teeming with railroad men and loggers? A brothel, that was it. That would make her richer than anything. Sable's brows knitted up again. No, she had no idea of how to do that, after all, she was a lady, a widow true, but a trollop never.

Sable made her way inhaling the brisk spring air like everyone else. The click clack of her high heels grated in her ears.

Sable's quick mind went over the fact that a brothel in a railroad town and a logging town was a golden opportunity. She also dismissed that thought because she knew that she did not have the temperament to put up with a bunch of sniveling whores. As she thought of this, a wry grin tugged at the corners of her mouth, when she remembered telling her mother one day that she had more respect for the women who sold their bodies than for those that gave them away. Her mother had been shocked, but Sable felt the sin was no worse, and only a fool would give away what she could sell. Her horror stricken step-mother could not understand this vulgar thinking.

Smoke-filled air was all there was to breathe today but this woman called Sable also breathed a big sigh of relief. She was in the United States, and she was going to spend the rest of her life here too. She knew full well that all her greedy in-laws up in the Province of Quebec, Canada could now go and kick sand down a rat hole. She smiled softly to herself when she thought of the most greedy two.

These were the Dupree sisters, her late husband's only two siblings, Elan and Josephine. Sable found these two pretentious bores little short of repugnant. They had wanted his farm, they had wanted everything that was his, but his widow, Sable Lore had gotten it all. As long as she stayed in the United States they could go to Hell.

The other reason Sable was so glad to be here was that

she could go and see her father and mother when she chose, but they were no longer close enough to have their noses in her business and chastise her every chance they could. Again a sly smile stole across Sable's face.

She had money from being a widow once. And now after going to the bank the morning after she buried her second husband, she had more, because she had put the newly inherited property up for sale. The banker in turn had found a quick buyer and in less than two weeks the deal was done. The farm had been sold lock, stock and barrel complete with household furnishings. Sable had gotten a good price for that working farm with over four hundred acres, barns, fine, healthy animals and well maintained tools. Andre's farm had been paid for by his father, years before he met Sable, but it didn't matter, it was hers all hers! Widowhood was lonely, but could be very profitable.

No doubt about it, Andre's snooty, spinster sisters would be most upset, and this brought a broad smile to her cupid's bow lips. Upsetting those two old bats was good. Now that Andre was gone they would have to go and freeload off someone else, unless the new owners would consent to letting them stay on in the apartment in one of the upstairs wings, as their brother had.

Andre was a nice looking man, but he had a nagging cough when Sable met him. He blamed this on his pipe-smoking, but it was not a smoking cough. Sable felt that it was more than that, and she had been just right too, because nine months after their wedding day the local doctor had confirmed her suspicions. He was afraid her husband was dying of quick consumption.

Sable had listened dry-eyed and calm to the words of the country doctor, and then she sat about making Andre comfortable in these his last days. At her urging, he obediently had gone to his attorney and executed a will. For his sisters, Andre had left all their mother's jewelry, a trust fund to take care of them, and a goodly sum of money to get them started somewhere else. This was followed by a Warranty

Deed with Isabel Lauren's name on it, making Sable the sole owner of his farm at his demise. Sable did not have long to wait, four weeks to the day Andre was gone and Sable was once again rich. A mischievous smile crept across her lips. She was free as a bird in a new country and ready to start a new life with not one, but two fortunes to spend. Life was sad, but on the other hand, life was good, and she just had to count her money and roll with the punches.

Andre's bank accounts were now hers and she added them to her own. Sable's blue eyes darkened when she thought about her sisters-in-law, they had never liked nor accepted her. They had always thought that their brother had married far beneath himself. They looked down at Sable feeling they were far above her, and Sable had known that from the very first. And so what if they were? To Hell with them both, thought Sable as she fought her way through the milling crowd.

She would never have to look at their disapproving faces again. Elan had the neck of a swan and the brains of a snail. Josephine was into fashion and had three closets full of expensive clothing, however, nothing she put on could hide those sharp features of hers or that nose that stuck out like a beak on some scrawny, strange species of bird.

Speaking of birds, both those old birds would be squawking for days and bad-mouthing, Sable, the "Scourge of Scotland,"

All those times before when Sable had heard them whispering about her in French, she hadn't given a damn what either of the gossiping old fools said, but somehow today, she would love to hear them!

Chapter Six

This was the twenty-fourth spring of Sable's life. She had been by then twice married and twice widowed. Andre's passing had been so quiet, compared to the dramatic demise of her first husband.

Her first husband had been her first mistake, or was it his, that Englishman who married her.

He was an artist and while vacationing in Canada, he had been commissioned by Sable's father to capture her likeness and paint a picture of Sable. She was just seventeen years old.

The dashing young artist, named Jonathan Glenhill, was captivated by his subject from the moment he first saw her. She was a true beauty, he thought with her reddish-gold locks, and the fire of green eyes. Jonathan was in his early thirties, and by the time he had completed the portrait he was madly in love with Sable.

Sable's parents were dead set against this relationship when they realized there was one because Jonathan was too old and their daughter was too young, but two months after the portrait was finished Jonathan returned to England, and Sable went with him as his wife.

Jonathan, that love-struck new groom lavished upon his new bride expensive jewelry and every possible luxury and comfort. He adored showing the beautiful young woman from Canada to his friends back home. She was his possession, and for three short months, he kept Sable happy with these material things and himself happily drunk, because Jonathan was not only wealthy he was also a drunkard.

He had not earned his fortune, but had inherited it from his dear departed mother. His life was filled with mood swings and he was known for going from melancholy to deep, dark depressions. He was what he had been for years, a manic depressive and Sable had kept his mind on other things for a little while, but now he spent more time in his studio with his bottle than he did at home with his wife.

He had had mental problems for years and Sable was about to find that out. With that discovery, the romance soon left this union and Sable decided that she had had quite enough of the marriage thing. She was disenchanted with her life, disappointed in her husband and bored with his snooty friends. And the day she decided that she didn't even like jolly old Jonathan himself was the day Sable announced that her new husband could take his bottle and go to Hell with it, because she was heading back to Canada. Her mind was made up. She was going home, with him or without him, preferably without the bore.

Their last days together were spent in the midst of horribly loud shouting matches, screaming fits of temper, verbal threats and broken furniture. The servants still called him "Master," but they called Sable that "little she devil."

The final brawl had taken place at Briarcroft Manner, Jonathan's family's summer place, and it was here in that elegant setting, overlooking the ocean, that Sable told her husband she was sailing for home that very day. Jonathan forbid her to do it, threatening to "cut her off without a dime," and Sable had told him, "Go ahead you sniveling bastard! It will be worth it not to have to look at your miserable face again!" These words crushed Jonathan.

The servants remembered the incident of Sable's departure very well. Jonathan had threatened that if she left him, he would hurl himself out the third story window onto the jagged rocks below, "I'll do it Sable. It will be your fault, on your damned head, but I'll do it. If you go out that door, I will jump out the God damned window," he had bellowed to the love of his life.

Those were the last words the servants ever heard their master say. The she devil had kicked a chair over and thrown a lamp at Jonathan and threatened, "You worthless drunk, if you don't jump, I'll push you!" And there were servants of that household who went to their graves believing that she did.

Without a backward glance, Sable had sailed for Canada, and Jonathan had finished his last few golden drops in the bottle, tossed it to the floor and eyed the window. His last earthly sound was a terrifying scream as gravity pulled his body to the cliffs below. This talented and troubled soul was finally free of the bottle, and of all his earthly cares. And yes, one rich young widow was finally free of him.

When Sable showed up with her two small suitcases many days later at her father's house, he was somewhat expecting her. He had already been informed of Jonathan's tragic passing through his attorney.

"Where is your husband, Sable?" He questioned.

Sable handed her luggage to two waiting maids and without a word started up the winding stairway in her father's sprawling home. Color rose on Robert Bigelow's face and he asked louder this time, "Sable, do you know where your husband is?"

"In Hell, I hope!" She exclaimed. She had turned midway up the stairs and faced her father with a steady, cold gaze. It was then he broke the news to her that she was a widow. Jonathan had committed suicide.

This news would have put most young women into a faint, but Sable just looked at him and said simply, "What?"

"I said your husband is dead, Sable. He committed suicide by jumping out a third story window while you high-tailed- it for home! Now, my young lady, you tell me what happened!"

Sable looked at her father, and her face was high with color and her green eyes were flashing and she answered, "We had problems. He drank far too much. We had a row and he said if I left him, he'd jump out the window to the rocks below and kill himself." Her father shook his head and when she said no more he questioned, "Sable, the man is dead! Is that all you have to say?"

Sable answered, "No, I've one more thing to say about Jonathan Glenhill. I'll bet it's the first time in his drunken life he kept his word."

With just a hint of a smile, and not a trace of remorse, Sable began climbing the stairway ahead of her, leaving her shocked step father alone with his questions. Her back was straight and she never gazed back once as she lightly stepped on the red carpeted stair treads. She headed for her own room and she knew that whatever the future brought, she was leaving whatever baggage that fell by the wayside, right where it belonged by the wayside and in her past another lifetime ago.

"Sometimes, I think she is a monster," whispered her mother. But the man whom Sable called "Pa Pa" knew that was not so. Sable was not a monster. She was a MacCutcheon, that's what she was.

Robert Bigelow went to his study then, to write a letter to inform the attorney handling Jonathan's affairs that it was with deepest regret that the Widow Glenhill had received the bad news. She was unable to write herself as she had taken to her bed, unable to handle or deal with any of it. Her father would do that. Robert explained that even though shocked and grief-stricken, Sable requested that any property left to her be sold and after expenses be placed in her private bank account in Montreal.

And now, years later Sable had gone through two husbands. Good old sickly Andre' had made it so easy for her. Her step-father didn't have to go through any rigmarole to get this money, because she had taken care of things and had it all with her.

Her heels made clicking sounds as she walked along

the wide boards that served as the platform of the Grand Trunk Depot.

Again she touched with ginger fingers the slender hatpin that secured her pretty green hat to her auburn hair. She smiled slightly as she saw the shocked look on the fellow who had been in the hooch, and whom she had been forced to use the hatpin on. He had screamed loudly and ran off like a scared cat, when Sable had smiled softly and sweetly said,

"Excuse me, Please," and then stabbed him in the upper arm so fast that he didn't know what was happening.

He had called over his shoulder that she was a "li'l, God Damned Witch," and if Sable had been, she would have turned him into a toad! She reasoned that he was just looking for a little excitement and she gave him some.

She knew that the young buck would think twice before he made unwanted advances toward another young woman traveling alone.

In her heart, in a small corner of it, Sable knew that what she had done was not very nice, but she did not intend to be nice, and neither did she intend to be a victim!

"Not in this lifetime, a victim of this toad, or any other," muttered our Sable under her breath.

Chapter Seven

Sable asked where the best place to stay in town was, and a kind, elderly gentlemen removed his hat and pointed across the street toward the hotel. The hotel incidentally, was not very impressive to Sable's way of thinking. It needed a coat of paint badly and there seemed to be nothing but lumbermen and river drivers of all sizes and shapes going in and out of the front door.

It's probably just a brothel, thought the small woman who clutched the one valise tightly, but the man who had given her directions seemed to be reading her thoughts. He assured her,

"Miss, I can assure you it is a respectable place."

Sable Lore thanked him kindly and moved on, her eyes never wavering from the establishment to which she had been advised to go.

Climbing up the wide board steps, she went inside and met some burly woodsmen coming out. They stopped and looked at her for one long moment because they enjoyed the soft scent of her cologne, and even the way her nose turned up at the very sight of them.

Sable did no acknowledging of their presence and not a man among them would have ever guessed that secretly she was enjoying the strong scents of pitch and pine and spruce and fir which adhered to their clothing. The lush, fragrance of the evergreen forest was everywhere, on their boots and even in their hair.

She breathed deeply and then lightly stepped out around them, thinking of their hair. She hoped that passing so close to them that she would not be visited by a friendly louse that jumped from their hair to own cleanly washed hair. She wondered if a louse could jump six feet. God, she hoped not.

With this unpleasant thought, her brows knitted together and she stepped to the front desk and pressed her small finger on the bell because no one was at the counter.

Sable's eyes traveled around the hotel. It was better on the inside that it was on the out. It looked and smelled clean, not fancy mind you, but it would probably be fine for a couple of nights, until she decided what she would do next.

The desk clerk appeared then from around a corner and was instructing a house maid to clean out a specific room thoroughly. Sable heard other voices then, and a huge man in a business suit and carrying a physician's black bag walked into the lobby. The other men called the big man "Doc," and were confirming that someone was dead.

"He has been dead a couple of days, judging by the looks and smell of him," the doctor person said. Sable stood quietly and listened at the same time thinking, "I hope they don't give me that room and that bed! I've had enough corpses in my life for awhile!"

The clerk, taking his place behind the front desk, greeted Sable with a slight but business-like smile. He peered at her through small, round wire glasses. He was of medium height, thin in stature and scarce of hair, making great contrast to the bearded, robust woodsmen prowling about.

"Ah, yes, the lady needs lodging?" He verbally guessed. Sable nodded her head, but looked sharply at him; of course the lady needed lodging. What in Hell did he think she had come here for, to count corpses as they dragged them out?

The clerk flipped the pages of the guest register before him, looked nervously over his shoulder and said,

"Well, this is our busy season, taking the drive down they are, you know, and most of our rooms are occupied now. We will have something after 6:00 p.m. er . . . ah, a gentlemen will be checking out."

Tired, impatient and extremely intuitive, Sable (whom was said to have the gift of the fey) said quickly, "Don't you really mean kind Sir, that he has already checked out, a day or two ago?"

While Sable tapped one small foot up and down on the floor waiting for his answer he said,

"Well, I'm sorry, Ma'am but that is all we have."

"What a pity! I won't be lodging here. I do not want that room nor that bed and if you have no other, no sooner, then my business with you Sir, is finished."

The desk clerk's eyes were small and beady but he was forced to hide a rye smile. And even as he did this, Sable was turning to leave the establishment.

"Hold on Miss," the stammering clerk made a waving motion and then she turned and looked at him, setting her small, bulging valise down with an impatient thump. The man continued,

"We do have one other possibility. We call it our bridal suite. It has a living room and a private bath along with a spacious bedroom. It is on the second floor and one has a good view of the river. This is as interesting time to watch the activity on the Connecticut River, where the tents of the river men can be seen all along the meadow. Most people find it very entertaining to see the burly river men as they take the log drive down."

"Who cares?" was Sable's question. "I asked for lodging not entertainment." This took the clerk back some. She apparently did not find this annual sight the least bit interesting, talk about entertaining.

Ignoring her words the clerk ventured on, "Of course, this suite being what it is, it will cost you more, quite a bit more, and most are not willing to pay for the comfort nor the luxury it offers."

"Neither, am I," Sable quipped, her Scottish roots again taking over her strong personality. She had enough cash in her valise to rent the damned hotel for a year or to buy it but she would be damned if she would pay a dime more than this damned "bridal suite" was worth.

As she mulled this over, the man suggested that they go and see the vacant accommodations before Sable made up her mind and she was pleased. She would take a good, long look at it and then she . . . not he . . . would decide how much luxury and comfort it provided.

She reiterated this and the clerk fetched a master key ring and led her up the long stairway and down a hall to the door. He opened it and they went in.

Inside, Sable saw a spacious living room with a roll top desk, couch and matching easy chair. The furniture was made of oak and the walls were covered with white paper with small green ivy scattered in a pattern that was subtle, tasteful and easy on the eye.

Sable walked to the bedroom and found the drapes and bed covering to be a solid green each in a velvet material. The rooms were light and pleasant but Sable decided to haggle. She no doubt would take the suite, but not at the price this bandit had mumbled on the way up the stairs.

Without a word she walked to the bathroom and her sharp eyes flew around quickly, darting here and there, just looking for a flaw, any one or two flaws that the eye of a man or even of an eagle might miss, but not the eye of a shrewd woman. In mere seconds she found what she was looking for and smiling softly, Sable pointed to the water-closet where she had spotted a tiny, hairline crack that could just barely be detected.

"This should be fixed before it becomes a problem," she complained and then she went on

"And, I see the bathroom door sticks, it is hard to shut and takes all my strength to move it."

The clerk's eyes shot up in complete surprise. This was the best they had, and it was apparently not good enough for this discerning lady.

Sable then became the charming and gentile woman that she was capable of being and she smiled, "Flawed as it is, I would be willing to rest here for a few days. I am a business woman, not a bride and I therefore do not need a bridal suite. The price you quoted me was for two anyway, I am sure and since I am only one, I find the price you quoted me far too much."

He held his eyes on her and repeated the price, and here Sable giggled and reminded the man,

"Sir, I want to rent it, not buy it!" The clerk was astonished.

Before he could sputter back Sable simply explained to him,

"I don't need a bridal suite. I'm not enchanted with watching a rough gang of men move logs and now what is your most reasonable fee for the adequate though some flawed accommodations?"

The clerk stood there staring at this obstinate creature in green velvet, wearing a demure smile. Until this day, he thought that the rudest person he had encountered while showing the suite was the arrogant George Van Dyke, of the Connecticut Valley Lumber Co., but this hellcat took the cake.

The clerk suddenly had an idea that this overbearing female was one of Van Dyke's relatives.

"Are you related to George Van Dyke?" he asked.

A complete blank look came over Sable's face and she raised her eyebrows and questioned,

"George Van Who?" And the desk clerk wished secretly that important old George could hear her say those words.

The clerk saw Sable didn't know the Van Dykes. He looked at her impudent face and said,

"We had not noticed the small things the lady brought to our attention. We will speak to maintenance about it immediately, and if you want the suite you can have a ten percent discount per night."

This clerk could speak as "We" because he spoke for himself and the other owner, his aging sister.

Sable wore a smile of mild satisfaction, because she hated to expend a lot of energy fighting when she could dicker. And she loved to dicker. She countered quickly,

"Nowhere near enough. I say, twenty-five percent discount," to which the clerk shook his head and said,

"That's robbery for such clean surroundings, not a bit more than twenty percent, and that is my final offer Madam."

A light girlish giggle floated up from Sable's throat when she agreed,

"I'll take it. It is clean, flawed but clean."

Still clutching her valise, Sable followed the clerk back down to the desk where she signed the register as Sable MacCutcheon.

She was so enchanting when she smiled. The clerk had noticed the large diamond and wide gold band on her ring finger, and he asked her cheerfully,

"Will your Mr. MacCutcheon be along anytime soon now Madam?"

Sable smiled and said, "Mr. MacCutcheon is dead. I am a widow."

Turning toward the stairway she had to wait until a litter was taken past her. Two men carried it and although the body was covered with a white sheet, there was no way of covering up the stench of death. The air reeked. Sable stepped back and sucked in her breath sharply.

The contrite desk clerk remarked solemnly,

"Well, Widow MacCutcheon, there is another widow out there tonight."

Picking up her valise Sable quipped,

"Fine, get her to sleep in the bed that one died in, for the Widow MacCutcheon sure as Hell will not! All respect kind sir to your generous first offer."

It took all of his professionalism to keep his voice level when he asked if she had any more luggage. To this she shook her head no, and proceeded to give him a smile so sweet that it could take one's breath away. The conniving clerk scratched his head. What a strange woman, indeed. He wondered if she really meant to be mean, or if she was just naturally mean?

Sable disappeared then in a swish and whirl of her long, green velvet skirt. Up the stairs she took her naturally-mean little self, leaving the man behind the desk feeling that he had been hit by a cyclone, but he wasn't sure from which direction.

Chapter Eight

The frogs and the peepers were singing with great voice on the meadow this night. Each one was telling his own story to the night. It was almost a happy sound and it should have been. It was well known in the north country that all was right with the world when the peepers sang. So in tune with all of nature were God's littlest creatures, it seemed strange that they did not know that these quiet men were still heartsick from the tragedy they had witnessed a few days prior. No, it seemed they did not know, nor did they care, those noisy critters so full of life and too loud to be ignored. They were part of life but they knew nothing about death. They didn't think about it, or worry about it any more than life, they only sang to the group of men gathered before their flickering camp fires.

This was a good balance because a river man knew a lot about death. It accompanied him each waking moment and it followed the drive with the doggedness of a camp following prostitute.

The lights of North Stratford were cheerful to see and about an hour after sunset many of the men began the trek across the river to that place where the lights burned, so that they could forget their sorrows by drowning them in rum. Women were there also, women both good and bad, each with their own special way of soothing.

"Is Big John goin' over?" Beaver questioned and the younger man nodded that he most likely was.

John was a bateauman and a good one. He was known for being able to maneuver a bateau in the roughest of white water. He was also known to have seldom ever been beaten in a fight. But, he was not a troublesome fellow, just one who could fight when he had to. They also said that they didn't make whiskey enough to get him drunk. But John had been drunk many times in his thirty-six years, only never while the drive was going down.

"Hell, he ain't goin' over to drink, just to fight and then carry one or two of them useless, young hellions home," Beaver said. And this was true. Big John had done that many times, the same as he would a bag of grain across each massive shoulder.

"I didn't think he'd come back this spring," Beaver said and spat a mouthful of tobacco juice on the ground.

"Probably wouldn't if his new wife hadn't died," Jake speculated and Beaver nodded his head in agreement and added,

"She was a flat-land woman, from Massachusetts, I think. Had a hard time adjusting to the country and the climate, got a cold that turned into quick consumption and died, as I heard it."

"Boredom, morn' likely killed her, "Jake said, and Beaver snickered. Eyup he guessed young Jake was ready for what ever North Stratford offered tonight.

Looking down from her window in North Stratford, Sable saw the rows of tents across the meadow. More lights came on as the night wore on, and the town was beginning to fill with burly rivermen. The prostitutes were mulling around in good number too. Their foolish giggles spreading upwards from the alley below where they walked with their men friends. Sable's first thought was, "Listen to those brainless bitches." And then her practical side came into view and she reasoned as she always had, that if a woman was that kind of a woman, it was smarter to sell her body than give it away. That was completely stupid!

Sable's head ached dully and had ever since she woke up from her nap. She realized that she was more than a little hungry and hadn't eaten a morsel since breakfast that morning.

She ran the comb through her curls, pinched her cheeks to make them rosy and then headed down the stairs to have some supper. On the way down the long stairway, she could smell steak frying and when she arrived at the hotel dining room she found it packed to the rafters with men and women, who seemed to be ravenously eating what she was enjoying the aroma of . . . rare steak.

Sable found a table in the back. She marched toward it and was about to sit down when her attention was drawn to a battle in progress going on right outside the window near the table she selected.

The other side of the glass, two men were pounding each other something fierce until one went down in the mud and the other one leaped upon him and beat his face without mercy. It looked like it was all over when suddenly the man on the bottom cursed loudly and gouged at the eyes of the one on top, and with that he rolled off and to the side, and the situation was immediately reversed. The man on the bottom was up on his feet now, and the other one was on the ground flat on his back. Blood was spurting and a woman standing near Sable screamed and fell into a faint. Sable stood frozen where she watched more men get involved in whatever had started this brawl. The fracas continued and got worse, when the man who had regained his feet began kicking his opponent in the head with his spiked boots. The gent with the gouged eye was blood covered and a tooth lay beside him.

More men and a couple of women were heading into the skirmish every minute but no one was in any hurry to break up the fight. Men were clearing out of the dining room and outside the window now was a pile of flailing hands and feet. Sable's eyebrows rose in surprise when she saw a large woman, carrying a big black umbrella circling the gladiators and beating the Jesus out of them with the parasol. The woman was middle aged, and as tall as many of the men in the brawl. She had red hair with a lot of gray in it, and she apparently didn't like anybody because she flailed men and women at random until she saw one fellow whom she recognized. He came crawling out of the mess, on his hands and knees and he was treated differently. She would rap him back of the head with the umbrella and then kick him in the behind with her black, peaked toed shoes. The crawling man made his way to a wagon and the warrior woman kicked his behind up into it. Then she got in, lashed the horses and they bolted out of sight.

What kind of a place was this? Sable was asking herself this question as she sat down. The crowd still milled around outside the window and now a couple of street women were into it, pulling hair, scratching, hissing like cats and rolling in the mud. Sable decided they were evenly matched and that they were not as agile, but fully as vicious toward one another as the men had been.

A serving girl came to pour Sable's water and give her a menu. Her eyes traveled to the window where Sable's gaze was still fixed. The waitress smiled,

"Pay them no mind, Miss. It's just those drunk and crazy log drivers again. And the trollops fight like tigers over them, you know." No, Sable didn't know and she really didn't care, but she quietly ordered the pot roast and hoped it would be well done. The blood outside had taken her desire for rare steak away.

Looking out the window once more, Sable saw that many of the men who had left the dining room had not actually gotten into the fight, but had only gone out where they could get a better look. Sable was thinking this sure was a place full of ruffians and barbarians, as she watched the cheering onlookers, encouraging the gladiators they had picked out. To her surprise, the women were the most vocal and two or three of the most prosperous and portly men were now in the middle of the fracas.

Then Sable saw another woman coming with an

umbrella and she was forced to laugh out loud when one bully of a man tried to wrench it from the female's hand only to find that quick as a wink she had turned the umbrella and hooked the handle around his ankle and down he went like a ton of bricks and then he really got a fair trouncing. A light, girlish giggle fluttered from Sable's throat then, she was thinking that one of these tough women wearing spiked boots and armed with a good sturdy club or parasol could probably tree half the tough, burly woodsmen in the north country.

Now, other town citizens were coming forward and appeared to be trying to break up the melee. Wasn't that always the way? Just when Sable would have considered placing a large bet on a couple of the combatants herself, they up and stop the damned fight! Was this Cud MacCutcheon's daughter or what?

"Excuse me Madam, but are both seats taken?" a low, masculine voice asked. Sable looked up and saw nothing, so she looked down and saw a dwarf in a fine suit standing beside the table.

"Of course one seat is taken. What am I, invisible?" Sable answered the little man and when he made a hearty laugh, all eyes turned toward them.

With his hat in his hands he bowed and rebuffed her then, "I assure you good lady, you are very visible, and I have no other intention but to eat. I only asked for the extra chair."

Sable consented and he took the chair and dragged it over to another table and after two great hops he was actually able to climb right up into it. Sable decided that she herself was built low to the ground but this poor creature was far worse. She was just beginning to feel a twinge of remorse, for not letting him eat at her table, when the waitress appeared with a plate of steaming pot roast, with browned potatoes and vegetables. Sable forgot about her less fortunate fellowman and began stuffing her face with both hands. She knew that a lady shouldn't eat ravenously nor a lot, but she would rather not think about this when she was so hungry that she could have eaten the south end of a north bound hog. Apple pie came with the meal, and Sable ate it with the same gusto with which she had eaten the pot roast. By God, thought Sable this town may be full of fighting fools but it also had some damned good cooks! Now Sable could use a good cook.

Sable finished her tea, looked around the dining room and rose to go back to her room. The crowd was still milling around outside the window, but the fight was over. Sable heard loud voices coming from the doorway. It seemed that two enemies of old had met there and neither would move out of the way to let the other pass. They growled lowly and then headed for the back of the building to settle the matter. A small group followed the two outside, but Sable did not go. She did not care who they were, what they were fighting over or who won, for that matter. Then her gaze fell on a tall, agile man in his mid thirties who had been unconsciously studying her and he had the audacity to look right at her and wink.

Sable walked past him, nose in the air and just as she got to the foot of the stairs, a middle aged man passed her dragging on an elderly white-haired women, who apparently had lost her senses.

She was wrinkled and bent but her eyes were like a bird's, sharp and they never wavered from Sable's face.

"Witch! That one — See, she is a Witch!" she screamed pointing a bony finger at Sable.

All eyes turned then on Sable. She held her head high and met the old woman's gaze with a look that would have frightened a women with her wits about her. It did not stop the screaming demon who was being dragged away up the stairs to her room. On the fifth stair she broke loose and came scampering back to where Sable stood frozen.

The elderly crone reminded Sable of a wizened up old squirrel in a black dress. Sable shook her head and listened to that accusing and shrill voice again. The voice of the squirrel woman got louder and more raspy and Sable knew she should have felt sorry for the crazy old coot, after all, it was plain to be seen that her aged accuser (all respect to her) had fallen out of an ugly tree and hit every damned branch.

"You don't fool me with your fiery curls and your good looks. You are a Witch, an evil Witch that should be run out of town or burned at the stake!"

Sable raised her hand high in an attempt to silence the woman and said low and controlled,

"Good Lady, you are mistaken or else you are mad!" The white haired one then shook her head as if to clear it, when she realized she had been called mad.

"Mad, tis Mad I am eh? I think not my young — Witch, Witch, Witch," she chanted as with the assistance of another guest, her middle aged son succeeded in dragging Sable's accuser away.

The shrillness of the older woman's voice had caused dead silence in every direction.

Sable could feel the red flush on her face and her temples throbbed with temper not embarrassment. Sable knew this was one of the most unpleasant experiences and embarrassing moments of her young life, but her anger prompted her to wish that she was a Witch, because she would snap her fingers and turn that menacing old bag of bones into a pile of goat dung!

There were people standing all around Sable and she could not move up the stairway. Then from the doorway emerged the giant who had dared wink at her earlier. He was wearing the attire of a woodsman, and a knowing smile. He took her arm and said softly,

"Might I escort the lady to the top of the stairs."

Sable cocked her head at this chivalrous one, who was the first apparent friend she had made in North Stratford. And, why in Hell did he want to befriend an accused witch? Was he a giant warlock? Ah, this depot town was indeed a strange place, with deranged squirrels wearing black dresses, warrior women, burly lumbermen who stomped on each other in spiked boots and now a protective warlock was added to the mix.

Some of the anger left Sable's face then and a softer look of amusement could be seen and the stormy eyes began to dance.

She smiled demurely and nodded and as he took her elbow to usher her along he smiled,

"Witches are rare here. I just wanted to meet one." At that very moment the dwarf came walking out of the dining room.

Sable shot back, "Be careful sir, he did too!" She motioned toward the dwarf.

"He walked me into this establishment and until he angered me, he was even taller than you!"

Big John began to laugh then and Sable went on her merry way up the long stairway. He decided that she was trying to warn him not to "mess with a witch." He was thinking that she must have come into town on the train from somewhere and that she must be going somewhere, somewhere far away where witches go. John allowed that she was very pretty, and there was something strange about her, but he didn't believe she flew into town on a broom.

He joined his cohorts then for some gambling and rum. John's thought now and then returned to the lady he left at the top of the stairs, a woman whose name he had not asked and would probably never know. He didn't think for one second that she was a witch, just a bewitcher of men with the power to smile and take a man's breath away or to snap her fingers and turn him into a toad.

Chapter Nine

Sable had the pleasure of listening to two more fights before she closed her eyes in sleep. She decided they probably were fighting in the alley below her window because they had better light there than to battle in the dark. She had heard and seen more fighting in North Stratford in the few hours she had been there than in all her life before. She didn't know if it was the fault of the townspeople or of the rivermen milling about, but one thing seemed for sure, that there were few enterprising men here. If there was, someone would utilize the talent and establish a boxing arena! In fact, she would do it herself if a woman was only allowed to do such a thing but that would be considered as scandalous as running a brothel. Damn, thought Sable, what an idiot society we live in.

The next morning she did stand in front of her window and watch the rivermen at work on the meadow across from her. She wondered how they could fight all night and work all day? There seemed to be hundreds of them, too many to count and all moving. She had no idea what they were doing, but everyone was busy doing something. Everywhere there were men and horses scattered about the meadows. "All the King's horses and all the King's men," I've seen them she mused with no idea these rivermen actually were breaking out high piles of logs called "landings," on the bank of the Nulhegan. This took physical strength and a keen eye. With their peaveys and cant dogs they picked and prodded and tried to free the great logs to be rolled onto the back of the mighty Connecticut where the Nulhegan flows into it. Looking at the landings above them where they prodded and picked with their tools, these rivermen knew that they were flirting with disaster and death almost continuously as they did so. If they picked and rolled the wrong log a great many of them could begin rolling with a thunder that brought dread into the heart of men scurrying out of harm's way.

Beaver had often thought that this mass of logs piled up and frozen together in unseen places was quite possibly the biggest and most dangerous puzzle in the world. With the deadly accuracy of a good cant dog man, he hooked and rolled a medium size log. No other log that he noticed moved so he continued to grunt and strain. He knew that it was said that the most dangerous job on the drive was that of the bateaumen. That was indeed a deadly business as they guided their craft through white, turbulent water to close a "boom" or to free up a log jam. It was a fact that more bateaumen drowned than any other group of rivermen, but old Beaver had seen men flattened out like pancakes, and at least two that were beheaded as swift as if a guillotine had done it, when a landing "hauled" or got to rolling loose down on the men working below. It was a grumbling sound that turned deadly and thunderous and it was a sound that Beaver listened for as he worked. He wasn't as young as he used to be to get out of the way, in fact, if truth be known, he needed a damned good running start.

In the middle of the night there had come a soft rain and the meadow under Beaver's spiked boots felt like a sponge. The high water had pushed logs that had been put into the water once already up on the verdant, green meadows of the Upper Coos. And there those great spruce logs lay like slender beached whales. These lay scattered about the meadows and they had to be picked up and pulled or "twitched" to the landings by horses and that's where the teamsters came in.

Young Jake was helping today with that task, and he didn't like it. He was working with a Frenchman named Jacques who cursed and swore and laid the reins to the backs of the horses whether they needed it or not. Jake had rather be breaking out landings up ahead with Beaver than here drag-assing along, playing harness boy and helper to that mean son of a bitch, and these were Jake's own thoughts.

Big John was with the "van" of the drive. The logs being driven down the river had already been launched onto the back of the mighty Connecticut from the directions of the four winds. They came from Indian Stream, Perry Stream and all their tributaries. Now the Nulhegan drive joined in and more logs would be amassed and waiting to join in from the Paul Stream area straight ahead and going down, down with the drive.

John hadn't gotten drunk last night. He didn't have a "big head" or hangover as many of the men did this morning and he was glad of that. He had soon learned that a river man needed his breakfast with him, not puked up on the river bank somewhere.

He was the vanguard of this drive, riding a log with the sure-footedness of a cat as together they rounded a curve on the Connecticut River, heading south. He was the epitome of what every little boy standing on the bank wanted to be. He might have been a brawling drunk last night, but today he was stone sober and reveling in the dangerous work he did. At one point, John waved his battered felt hat to the onlookers that lined the shores. He was as arrogant as he was graceful. He seemed almost relaxed because he was so confident, but this man was alert. He knew more than anyone that alert men lived to finish the drive and careless men died. John's big, calloused hands wrapped securely around his peavey handle. His knees were slightly bent and his balance was uncanny leaning into the wind when he needed to, while his spiked boots dug into of the bark of the great log upon which he rode. He could scamper up and down that log like a squirrel, his eyes ever watchful of the bumping logs all around him. He poked and guided them along with his peavey, keeping them from snagging or jamming, making sure they did as he directed them to do, follow the river, ride the rapids and keep the drive straight ahead and going down.

Behind him moved millions of feet of virgin logs that were massed at Second Connecticut lake and these had shot down the roaring, raging river and were caught in the booms and held securely there, waiting for millions more feet of logs to join in. A part of the great northern spruce forests was floating along behind him. The logs would push him and shove him, and were sometimes determined to go by him, but he would not let them. He was in control here, he and all the other "Cowboys of the Connecticut," who did their ballet on heaving logs and moved the mountains to the sea. Did these log drivers show-off a little? No they did it a lot.

It was mid morning and the crew assembled here were the front or "vanguard," of this drive. They worked, great muscles rippling under woolen shirts. A great army of men who looked like ants from afar, that never seemed to sleep.

The day wore on and the sun rose higher in the sky. The rays were strong upon the human eye and made men blink. Jake shaded his eyes with his hand and strained to look ahead. The miles were passing and the drive was going at a good pace but suddenly up ahead the glint of sunlight upon ice caught Jake's attention.

The drive had encountered its first serious ice jam up ahead and thousands and thousands of thundering logs racing towards it. And there was no power on earth to stop the confrontation. Danger flags were being raised while rivermen called to one another. Great logs were hitting chunks of ice and hurling men through the air like they were sparks from a night fire. Men danced from log to log, bellowing, cursing and using their cant dogs and trying to help fallen comrades who now were trying to escape being crushed between great blocks of ice and mighty spruce logs.

The water was cold, so cold that it numbed the body and when the men came crawling up on a log or staggered toward the shore, they looked blue.

They tried in vain to stop the log jam but they could not. They could only scamper like the human squirrels they were and get back as one great log smashed into another, often sending them up in the air, end on end and the pileup continued to grow higher until it reached nearly forty feet high with logs sticking out in every direction.

Some said nothing could stop the drive from going down, but that was not true. The great battle between ice and timber on the Connecticut River was over this time, and the ice had won.

All along the shore, the men looked like drowned rats and extended helping hands to one another. Already danger flags were waving to warn the drivers coming down of the danger that lay ahead.

The men now were removing wet clothing, but not their spiked boots. They wrapped themselves in woolen blankets that usually felt scratchy as hell to bare skin, but right now felt wonderful to shaking, shivering blue colored bodies.

On the river, the groan and thumping sound of moving timber crashing into still timber could be heard, but on the shore the sound of chattering teeth was overcoming that of cursing.

"God Damn, the God Damned luck!" was repeated and then came a lament,

"JJJJesus H. CCCChrist! I've lost mmmy new ssstore bought tttteeth!"

"Lucky twasn't your family jewels," some unsympathetic river man said and a slight rumble of laughter was heard as these shivering men counted noses and decided they were all there.

"Hell," someone laughed, "Sam, you know you left your family jewels at a whorehouse back in Stratford,"

And Sam, that one who stuttered badly and who had lost his new choppers added,

"DDDamned rrrright an if I'd been ssssmart I'd a left my GGGod Damned Tttteeth there too!"

"Hindsight Sam! There is a Connecticut River Salmon wearin' them teeth now!" By this time the teeth fiasco was a big joke and the men roared with laughter. They had all narrowly escaped the jaws of death, and they were happy with that and had little sympathy for Sam's loss. This was a celebration of lives saved not teeth lost. Besides, some agreed with old Beaver who said,

"Hell, Sam couldn't talk a bit better with 'em nohow and they probably fit the salmon better."

There was not a man of them who did not know that breaking this jam was going to be a hell of a job, dangerous and mean. But break it up they would. That was what log drivers did. It was what they got paid for and they were paid well. Come Hell or high water, ice jams or log jams, with teeth or without them, they would take the drive down.

Chapter Ten

The rain and drizzle of the night before had cleared and the morning sun peeked through the heavy sky and gave promise of warmth on this the tenth day of May. The "van" of the drive had departed out of sight around the last bend in the river leaving North Stratford, that busy depot town behind in its wake.

Short days before, many of these river men had labored shirtless, waist-deep in water and had masterfully guided logs from rollways along the banks and now they had them calmly floating along on the fast moving waters of the Connecticut River. The logs, those heavy and cumbersome things had not all wanted to make this journey to Mt. Tom and only the persistent and constant prodding and prying of the cant dog and the peavey in the hands of experienced log drivers had changed this stubborn, northern timber's mind.

Up ahead the great logs comprising the front or van of the drive bobbed and floated along on smooth water. From the mountains where they had been cut and hauled to steep landings, millions of feet of logs bumped up against one another. And at times, in this moving but not turbulent water, men flitted across the logs driving, alert and tending out. The river men were always looking for the bad place up ahead. In these calm stretches of water, where the drive went peacefully down, there seemed to be a strange and peaceful acceptance by the logs. This is when the river man could strut his stuff, roll with the logs and lean on his peavey and enjoy this ride. But this strange peace between timber and men and this river was only a temporary situation.

This spring drive seemed to be in the full control of the river men. That great army of ants had been driving the logs, corralling them in great log booms hitched together with chains. These places where logs were held in wait and allowed to go when the men opened and closed the booms or gates was just one more proof that man was in charge. And so it seemed, these log driving men controlled the logs and the river. The invention of the squirt dams enabled great logs to be driven or sluiced through narrow places they could not have otherwise gone unless carried along by man's ingenuity and a great wall of water.

On the quiet places on the river, no river man was fooled into a false sense of security. After all, most of the bad places lay up ahead. And what lay behind were the remains of what had been the great standing spruce forests of northern New England, and now they tore down that black river with a vengeance as if to tell the log drivers,

"You bastards, you uprooted me from my mountains and you want to drive me to the sea, and If you think I started hard . . . try to stop me! Well, here I come, ready or not!"

Down the "Long" (Connecticut) River went the timber and the men. The drive from the several branches of the Nulhegan, Perry Stream and Indian Stream had been rolled onto the back of the river and was straight ahead and going down.

In the township of Brunswick the drive passed the springs. Most of the men looked up at it and saw the huge hotel rising there. It was an unusual place for here, just feet from each other were seven different kinds of mineral water. This was said to be a place of healing and the Abenakis called it the "Nebesonbek," or Mineral Waters. They drank of these waters and if they were injured let the healing waters run over their bodies, and they were healed.

One evening a few of the men swam across to look the place over. There were some Abenakis and Mohawk men on the bateaux crews. The Abenaki men wanted to go and drink some of the water and ask for healing anywhere they might be hurting. The Mohawk men wanted to go and watch this happen to the Abenaki men. Most of the White men on the drive wouldn't bother to go but Rene and Gaston, two French Canadians decided that they would also go and watch the healing process. After all, there were Catholic Shrines well known for miracles.

It was not the best night for swimming, but that was what it took to get to the other side of the Connecticut River so one-by-one the men dove in and swam across that cold, black water. Wet and looking like musk rats they crawled out at the foot of the steep bank leading to the famous springs. Gaston was the first once across.

"Don't drag you Hass's!" he laughingly called back to the rest of the men strung out behind him as he headed for the brush above.

Gaston was such a good swimmer that he was almost a minute ahead of the rest of the men to set feet on the opposite shore. He was boisterous, and had a tendency to say his sentences backwards if he became excited . The men would howl with mirth over his use or misuse of the English language, but he never got angry or insulted when the men laughed at his broken English. On shore, the crew shook themselves and started scampering up the bank following Gaston who was nearly to the top of the high embankment.

Gaston made his way along through the tangle of small bushes. Then he heard a grunting noise ahead and as he neared the top of the precipice he came upon a large female bear, clawing for a feed of ants from an ancient stump. She turned sharply and looked Gaston straight in the eye. She was startled and she stood up straight on her hind legs and then grunted and sat right back down where she was, flat on her big bear's behind. And, even though Gaston had stopped dead in his tracks at the sight of her, she was irritated by his company and she partially opened her mouth making a low throaty growl. Then she opened her great jaws and Gaston saw the dull yellow of sharp teeth and then still sitting on her butt she made a terrible growl and seemed to be grinding her teeth at him.

Rene and the others were making their way up the steep incline when they met Gaston coming back on the dead run, headed for the river.

"Run, Run for you life! Me, Hi just saw dat She bar!" The men were snorting out now at the sight of the little Frenchman making tracks for the water.

"Come back, Gaston," called his friend Rene. "She go haway, dat She Bar!"

Gaston turned and yelled back to his advancing comrades,

"Ho, Yeah. You tink so! Well she right ahead of you! She sat right hup on her big teeth and she grind her Hass at me!"

It was with blood, sweat, tears and laughter, that the longest log drive in the world, pressed on, leaving the beautiful and verdant meadows of the Upper Coos in the wake of thundering logs. And that incredible passage left something burned forever into the memories of the magnificent men who rode them down.

Chapter Eleven

The logs from First Connecticut Lake in Pittsburg, N.H. were leading the way and setting the pace for all that came after them. The rivermen who joined in where the Nulhegan flows into the Connecticut River had described that drive as "one bitch of a thing!" But this was the very nature of the Nulhegan. This river was like that steep and winding valley itself, having everything to devil a log driver and prevent him from doing what he had been hired to do, drive logs. This was a river with several different branches that were logged at various intervals, but on the East Branch of the Nulhegan there were places that were just impossible to ride a log. You could drive one, but you wouldn't ride one!

During this season of high water, right after the ice went out this river became a roaring flume with walls of rock on the sides and on the bottom too. At higher elevations up on steeper ground numerous "squirt dams" had been built in strategic places that provided water pressure enough to flush the logs down stream. Down the narrow, winding and deadly places those logs went like a bat out of Hell. A roaring and great wall of water was pushing them and they were all pushing each other. And ahead they tore faster and faster down ravines so steep and winding that even those fast moving logs had to be directed and forced around every protruding rock, tree and bend in that river. When crooked logs weren't doing their best to catch on something and cause a jam they were famous for ramming themselves helter-skelter onto the nearest sand bar, and the Nulhegan like all the other driving streams in the area had plenty of those.

Any log driver who ever stood in the morning mists a little before daybreak and went about the business of "tending out," or keeping the drive progressing, certainly had his work cut out for him. The men rose from their sleeping blankets before the sun rose, and after a breakfast that would have sunk a battle ship they were expected to go out and dance a jig on a log.

Those tough and magic men were often ghostly figures working in or near the river, wrapped in a blanket of fog so thick they could not see each other. The swirling mists of early morning hid the men from their work and in fact, often times, the men working with their cant dogs couldn't even see the river itself. They could not see it, but to a man, they could feel its dampness, smell the muddy odor of it, and they could hear the muted gurgle of it, as it snaked menacingly through the meadows and the alder swamps that dotted the Connecticut River Valley.

Every part of log driving had a sound, and no two were quite alike. Even the great melting chunks of ice that floated past the logs and the men standing bare to the waist in that frigid water had a distinct sound when they hit a log. It was not the "thunk thunk" sound of a log hitting against a log, this was a log hitting a cake of ice and it made a grating sound, not a thunking one.

The Nulhegan Drive had been rolled from the rollways and was on the water and headed down. They were making good time, but up ahead above the mill town of Groveton, there was always an ice jam waiting to create a log jam on the Craggy meadows.

This was also a place where the Connecticut Valley Lumber Company usually had logs scattered all about these meadows, and Lumber King George Van Dyke was known for promising to pay the damages, but never seemed to get around to it.

The men sympathized with the farmers for having their fields dug up by the stranded logs, because most of these rivermen were also farmers, but God Damn it, the Drive had to go down! And, if a few angry farmers and their accompanying sheriffs had to be thrown into the Connecticut River to continue the plan, then into the river the farmers and their protecting law officers went. However, when the drive was done, had the bellowing boss of the Connecticut Valley river hogs paid his men with promises and not with United States currency his drive would have gone to Hell - not down. But the Connecticut Valley log drivers were loyal to old George. If he paid them to ride logs they would, and if he paid them to put down their cant dogs and peaveys and go to fist fighting to drive off anyone who disrupted his drive, they would. And many were the times they did. And nearly a century later they would recall those glorious days and have many great laughs over them.

For the bateaumen this drive so far had not been bad. Only twice had they had to go out into the Connecticut River and free log jams. They had freed them too, those mighty men who balanced those big boats and picked with their peaveys and rolled with their cant dogs and found the culprit king log or key log. These two jams were backbreaking but were relatively easy ones, because when the bad ones came into view the bateaumen carried dynamite attached to long poles. Armed with these trusty poles they set off in their boats rowing out to the great pile of logs that looked like thousands of giant match sticks sticking out in every direction. The skillful bateaumen brought their boats in close to the jam, all the while looking for the right hole in that tangled mess of lumber, into which they would stuff the dynamite. This hopefully would break loose the obstruction so the drive could continue on down. They wanted to get the drive moving, they wanted to hurt or damage as little wood as possible, and they also tried not to kill themselves or each other. However, dynamite sometimes has as strong a mind as timber. They said if it was frozen, you could thaw it out in an oven and it wouldn't explode, but then it had been seen to ignite and kill and cripple good men because it wasn't supposed to do what it did.

A Mohawk bateauman called Peter Jay who ran the bateaus with Big John had two fingers missing from his left hand caused by an accident that wasn't supposed to happen. It had happened though just two years ago. And Peter Jay had finished the trip to the sawmill at Mt. Tom, Holyoke, Mass. with his injured hand wrapped in great bandages. As the newer men came on, they would ask Jay what happened to his fingers and he would look very solemn and say,

"Dynamite him go up. Old Jay him come down. Fingers him land in Turkey!" He just didn't bother to say that no one knew why that fuse went off almost the second they lit it, because Old Jay he liked the "Turkey Story" better. He liked it better, because he didn't care to hear those young fellows say, "Oh, that's terrible. How do you get along without those fingers?" Besides, Peter Jay liked the way Beaver summed the situation up the day of the accident. After looking diligently but in vain for any sign of the severed fingers Beaver spat tobacco juice, passed gas loudly, and to the world proclaimed,

"They's gone Peter Jay! Gone to Hell, but ifn' a man's got 'nuff fingers left to scratch his behind and to pick his nose, he's got 'nuff. And you got plenty!"

Chapter Twelve

Joined or followed by the winter cuts from Willard Stream, the Mohawk, Simms Stream, the Nulhegan and Blodgett Brook, the main drive proceeded in a southerly direction.

The Nulhegan drive at Bloomfield, and the winter cut from Paul's Stream up in Maidstone, added a tremendous amount of wood to the timber already floating on the back of the Connecticut River.

Flagmen and flagboys were positioned on high points above a stream, and had the job of watching the logs and running a flag up a pole to signal to the drivers below. This code system was what directed the opening and closing of boom and dam gates.

There had been just enough rain so that the mosquitoes and black flies deviled the men whether waking or trying to sleep. The smell of citronella was everywhere along the meadow, mixed with the order of bodily sweat. The citronella helped some, but when the weather was just right the black flies and the dreaded no-seeums feasted on rivermen every hour of every day.

Since the drive had left North Stratford, the weather had cleared more each day and the past two days were so hot that some of the men worked without shirts. Most did not, because they could not stand the pain inflicted upon them by the buzzing insects and the intense rays of the sun.

Tonight the Maidstone Meadows above the town of Guildhall, Vermont was where the tents appeared. The drive still made progress each day with only miner problems.

The wangan train would soon be coming into view and this was an impressive sight to see. The wagons were long and drawn by sometimes four but usually six fine horses. The long wagons were painted blue and had CVL painted on their sides. The horses wore fine harnesses, bedecked with bright brass and even tassels. These wagons that made up the wangan were an impressive caravan to see, and they carried it all. Every necessity to supply the needs of the men and animals working on the drive was carried along the river by the wangan and these were loaded with food, tents, stoves, baking sheets, cant dogs, pike poles, chains and almost everything else including the all important smoking and chewing tobacco.

Once it began the drive never stopped and neither did the wagons.

Gathered around their evening fire the night before, several of the men who were locals from the Guildhall and Maidstone area, were talking about the old haunted house near the village of Guildhall. This was not far from here and it stood where it always did, looking gray and bent and lonesome even in the light of the brightest day. All the people this end of the valley knew that strange things happened there. No one had lived there for years but sounds were heard by passersby and there were many who testified to the fact that bright lights resembling leaping flames could be seen leaping throughout the house on any dark night. It leaned a little and sat right on the bank of the Vermont side and it was said that although the doors were never locked, no one ever went inside.

"By Tunkit, I don't believe in no damned Ghosts," announced Beaver and before too long a small chorus joined in agreement. Then Big John, who rubbed horse liniment on his sore and aching muscles asked,

"Well, has anyone ever stayed there to find out if that old place really is haunted?"

The consensus was that nobody had, but that if it was known to be a place where strange things happened, then it was known to be strange or "haunted."

Peter Jay sat quietly puffing on his pipe. He was a man who did not believe in ghosts but who did believe in the presence of the super natural. To Jay's way of thinking, he could understand a free roaming spirit, but why would any spirit wish to be a ghost and be confined to live inside a lonely old house when he had the whole forest to roam freely in?

Jake sat listening in the glow of the campfire. This was his favorite time of day, sitting by a comfortable night fire and listening to tales of other times and places.

Then someone suggested that some of the rivermen go over and spread their blankets inside the old haunted house, just to see what would happen. This seemed like a wonderful idea and without much talk Jake, Beaver, Big John, Bill and Peter Jay gathered their bedrolls and began the long trek to the haunted house.

Is was almost dark when they started out and it was completely dark when the five rivermen arrived. Inside they found an old lamp and with fumbling fingers Jake succeeded in getting it lit.

It added some light to the lantern they brought with them, but upon entering the house one of John's feet had gone through the porch but he pulled it out of the broken boards and ventured on inside.

With the eerie shadows thrown by the kerosine lamp and battered lantern all eyes were squinting to see what there was to be seen. There was a rickety table and two chairs in equally poor shape and a large wood cook stove. The lamp provided poor lighting to see much of anything, but with the exception of the hedgehog they had run into on their way inside the house, there seemed to be nothing too strange about the place. It did however, creak and groan with each move anyone made.

The ghost hunters took the lamp and made the journey up the groaning stairs and checked out a narrow hall that was strung with spider webs. There were two bedrooms at the end of the hall and these seemed to be bare of furniture, but under their feet the rustle of last year's autumn leaves that had blown in from somewhere could be heard. Beaver hacked and coughed from the dust but it didn't seem to have any effect on the rest of the crew.

The dirty windows were strung with cobwebs and many panes of glass were missing. The adventurous rivermen could look out and see the glow of the campfires across the meadow from which they had just come and they looked cheerful compared to this damp and lonely house.

Each man picked a spot and threw his blankets down and prepared to wait in silence for either sleep or ghosts to overtake them.

Only the gentle sigh of the night wind could be heard. Beaver went to say something to Big John but by the snore knew that John had gone to sleep already.

Peter Jay was the last of them to close his eyes. He was listening to the wind and wondering what had happened in this house to make the hair stand up on his neck like it had since he had arrived.

Sometime, long after midnight when all the rivermen were deep in sleep and the moon hung low over the northern Connecticut River Valley, a strange sound was heard.

One-by-one they heard it and rolled out of their blankets on the board floor. Was it a whine or was it a cat crying?

As the men came fully awake it was plain that the sound was neither a whine nor the sound a cat makes, it was the unmistakable sound of a baby crying. The house was still dark except for the moonlight shining through the dirty windows and the crying sound was getting louder now and was coming from the downstairs.

They headed down the stairs, their eyes still swollen from sleep, and were wondering how in hell their buddies over on the meadow had gotten a baby down there to scare the Hell out of them.

As the group converged on the kitchen there was nothing to be seen and after much groping around in the dim light of the lamp, it was agreed that the baby was crying louder all the time.

"It's coming from the cellar," said young Jake and immediately they opened a narrow door and proceeded to go down into that damp place.

Even the lamp didn't help much down here and the smell of the earth only reminded them what most of them were thinking, that there was someone buried down here, and that is why the house was haunted.

"We ain't got no shovel to dig up any bones," complained old Beaver.

"How do you know we're looking for bones, Beaver?" Jake wanted to know and Big John snapped, "We're looking for a baby, a crying baby, not any Christly bones!"

Peter Jay said nothing but he knew that there was nothing so bad as disturbing the spirit of the dead. Even if that spirit of the crying baby wasn't angry, its parents would be!

Beaver dug with one of his hands in the dirt of the cellar floor, but the earth was too hard and he had to stop. They kicked the cellar dirt with their feet and dug with an old stick they found there but they stopped digging because down in the cellar the crying sound was very faint.

Bill said, "Now that makes sense to me that someone had a kid that died and they snuck down here and buried it so as nobody would know. And that's why the ghost won't leave. It wants a decent burial."

Peter Jay was listening while all the rest were talking, and he held up his hand and said

"Baby not cry down here, him cry up there," and he pointed up towards the kitchen from which they had come.

"You're right Peter, there is no crying here, its coming from up there!" Big John stated this firmly and the procession started up the cellar stairs and by listening they could tell exactly where the crying was coming from — the big wood cook stove.

No fire had been laid into that stove for a long time but as they arrived back in the kitchen the crying began again.

They removed the burners and peered down inside the stove, holding the lamp at different angles so they could see what was crying or what had been crying, but the stove was almost bare of ashes.

Strange glances then fell across the faces of the rivermen. There was most definitely a baby crying and the sound was coming from that ancient stove. They rapped on the stovepipe, then took it down and tried looking into the chimney. They dismantled the entire stove until it was nothing but a pile of iron and all the while the baby cried.

Conversation was nil as the last remaining pieces of the stove were taken apart, and the rivermen had worked at their task all night.

Just as they first rays of the morning sun came through the dingy window panes a strange thing happened, the baby stopped crying. And one-by-one the men filed out of that strange old house and straggled across the meadow towards the river to begin a day's work. Not a one of them ever said that they were afraid, not a one of them ever could explain what happened, but everyone of them knew what they had heard.

That evening in front of their night fires they spoke of it often. And each man who was there and the ones who were listening to the incredible tale had their own idea of the mysterious goings on.

Beaver spat tobacco juice and put the whole incident into perspective.

"Well, twarn't no cat, and twarn't no wind sound. I was there and I heard it. I'm tired of hearin' it coulda been this or it coulda been that, 'cause it coulda been a God damned Hookadoo, for all I know, I didn't see it. But, I'm tellin' ya what I heard and by Jesus, what I heard was a baby cryin'!"

The men shifted around the fires. There was nothing more to say.

And so the tale went on with no one ever figuring out the mystery. The five men who spent the night with the crying baby never knew what happened, only that it did.

It was, however, told around the valley that many other unexplained things had been heard and seen before in that old house in Guildhall, which was said to be haunted.

Chapter Thirteen

This had been a somewhat uneventful drive so far. Not like the one nine years ago in the spring of 1890, when on the 8th day of May a tremendous ice jam took the bridge out on the Upper Ammonoosuc in the town of Groveton, N.H.

Big John was now in the prime of his life and had realized part of his secret ambition. He could ride a heaving log through raging white water with the ease and grace of all those other cat- like men who had driven the river before him, the ones whom he watched and admired when he was just a small boy standing on the Stratford Bridge. In his sprint from boyhood to manhood, most of those heroes were gone, but they had left their mark upon this man named John, and now he was that carefree, roust-about dancing on a log, and showing off with all of his good natured heart. Just as they were then, he was now.

He had given up much of his drinking and whoring around and he had kept his wedding vows, but when his wife died, it seemed that life had thrown him back to the waiting arms of the river, the "Gwenitegok" (Long Tidal River) who waited for no man. The Connecticut River had whispered his name and John had heard the call. This beautiful river snaked its way through the lush, green valley singing to the rivermen in some places and killing them swiftly in others. And the log drivers loved her more than they feared her or they could never have floated forty-five to fifty-five million feet of logs down river to the sawmills and paper mills every spring for half a century.

The days were spent driving the logs down river. Picking and prodding and rolling logs and clearing a path for the thousands of logs coming behind these in the van of the drive. And always the men were alert and looking ahead for the bad situation that could create a log jam and the Connecticut had seen them all. Great logs piled like matches were backed up for two miles or more. Some could be freed by the ingenuity of the bateau men and the river men who stood on a moving log and kept other logs from stopping the endless parade of wood. The very worst jams had to be blasted apart with sticks of dynamite, and these rivermen had the expertise to do that too. When you could dart across a river of rolling logs and toss a stick of dynamite at a strategic place that would break a jam and time it so that you were skipping back across other logs and out of harm's way, you could do something right. And you had better, or you were dead. There was no place on the drive for too many miscalculations, ask Peter Jay, who used to say the timing of the explosion that got him was only off by two fingers.

The "bad places" on this drive were up ahead and coming fast. The older, more experienced river men told the younger ones all about the dreaded Fifteen Mile Falls that began at Fitzdale (Gilman) in the town of Lunenburg and went over saber-toothed rocks and rapids for the next fifteen miles. You did not run these falls with a bateau and you did not ride a log down this dangerous stretch of water. Those who had tried, had died.

Jake stood on the bank mesmerized at the skill and ease with which Big John and Peter Jay pried and poked and finally freed-up what would have most certainly been a jam. The logs stuck out like quills on the back of porcupine, but these two seemed to know just which way to swing the bateau, how close or not too close to go, and then just which particular logs to poke at so that the problem would be resolved, not become bigger. They flirted with danger and death every day from the long bateaus they worked from, and they loved what they were doing, and Jake loved to watch them. He wanted to be just like them, cocky, arrogant, and proud of what he did. And he, like them, would get much personal satisfaction knowing that without men like themselves, the drive would not go down.

The town of Lancaster already lay in the wake of the drive. Some of the men talked about sneaking over to town at night to scare up some women and rum but by nightfall were too tired to go looking for either of these things. Lancaster was a fine town with a fine jail that through the years had seen its share of "happy or unhappy log drivers."

Some of the men said the boss of the drive, lumber baron George Van Dyke had a fine house here, others said it wasn't any better than a good many other fine houses in Lancaster and yet, all of them wondered how a river man, which Van Dkye was first, ever rose to what he was? He was their boss, and he also was living proof of all that they might be. Van Dyke was the undisputed King of the Connecticut Valley Log Drives, and though they knew him to be an unscrupulous, arrogant ass, they would have followed that old river pirate anywhere, and they did.

For several nights now, as they sat before their evening fires, the older men told the younger men to be very cautious, because the most dreaded water on the whole Connecticut River was straight ahead between Logan's Rips at Fitzdale and the dreaded falls a few miles below that were notorious and deadly and blamed for twenty deaths. And if you'd asked a river man the most dangerous spot on the Connecticut, he would have said that it was Mulliken's Pitch and the Seven Islands on the beds of gravel at the lower end of the falls. Many times the men who were drowned and killed here were buried in pork barrels. And as grim as this fate seems, it was a well known fact to a river man that it didn't matter a damn what you were buried in as long as you had died well.

John was a renowned bateauman and Jake followed him at a distance, seldom talking but always observing his idol. Jake too had a secret ambition, he wanted to be able to steer a bateau down the rough and dangerous stretches of river coming up ahead. He wanted to be just like Big John fearlessly enjoying each death defying day he lived, and walking on the water.

The sunset the evening before had been a spectacular one, and told of hot dry weather ahead. And this day had dawned and that fiery ball in the sky had a long way to go before it was directly overhead, and already sweat glistened on the bare backs of the men laboring under its scorching rays.

The wangan had provided all the workers with a good breakfast as usual of eggs cooked just right in a pan with bacon drippings spooned over them, toast, donuts, muffins, pancakes, hash brown potatoes and tea, coffee or milk. This was the time of day to eat, and brother they did, which is the second reason they called themselves "river hogs."

Most any man would tell you that when the sun rose and it got really hot and humid that few felt like eating a lot no matter how hard they worked. But in the morning when they were fresh, they could eat with the same gusto that they faced the day ahead of them with.

The Indian, Peter Jay said the Indian ate only what he needed to sustain himself, but the white man ate more than he needed habitually because in his quest for power,

"If him can't own it, him eat it!"

The philosophy of the log drivers always made sense and kept life simple. Beaver always claimed that life itself was simple until man complicated it with his laws and rituals and his schemes to control the earth and everything that dared to live on it.

The morning was moving along and the sun was racing across the heavens at a faster pace than the logs were thundering down the river.

Jake sat in the second bateau. He had done what he was told, and would do no less, and he was proud of what these rivermen had accomplished. Up ahead rode the brazen rivermen, their spike boots driven into a log, and balancing with the peavey they rode those rolling logs through that churning white water and around the sharp bends ahead. They did this with the same self-confidence with which they had ridden the logs through the narrow gorges behind them.

Sport for the rivermen was often that of "birling" or log rolling. They made great contests of this, where two men got on the same log. Each then tried to get the other guy to lose his footing and fall into the river, but they could not touch each other in any way. More experienced rivermen would rotate the log and do so faster and faster until the other fellow couldn't keep up, but if they were evenly matched the log would whirl one way then stop abruptly, start the other way, switch again until only one was still or left standing on the log. This took extraordinary balance and the quickness of a cat, because all that skill and more would be needed by the time they came to see what little a surging log meant to the beginning of Fifteen Mile Falls at Fitzdale.

There was a mill here, a small one, and that was the forewarning that where there was a mill there was a damned "dam" to contend with. The long logs would have to be sluiced or forced through the dam and on a head of water where they would be tossed in the air and hopefully not kick back, but fly forward and down leading the way for the millions of feet of thundering timber coming up behind . This was a likely place for a log jam and there were a good many here at this dangerous place.

Sometimes the logs went along real easy and then there were other times when they hit with a deadly, thundering sound and were stuck up into the air like pins sticking out of a pin cushion. As a jam was building the men signaled each other with caution flags and then there was nothing to do but to boom some of it, hold it back until they could go down there with their peaveys and cant dogs and pry, roll and lift. And now the rivermen were moving fast, literally racing across the top of logs as fast as they could in a desperate race to keep up with the timber. Because the rules of the game had just changed, and from the time the first log pitched over the dam and fell toward the river below, past jagged rocks and foaming white water, all knew the song the timber sang,

"Rivermen, you drove us here . . . now, you just try and stop us!"

Stop millions of feet of logs, of course, it really couldn't be done, but those rivermen would go down into the middle of that jam and pick and pry and break it loose. They would strain great muscles in backs and in biceps and they would labor and sweat and some might die, but they would never stop their pursuit. And they would keep the drive moving straight ahead and going down.

The thunderous roaring sound of uncounted logs being sluiced over this dam was ominous and everything stopped while this was happening. People lined the banks of the river to watch a sight they never got sick of, timber being catapulted into the air like bullets and sent down to the river below. It was with a deafening roar that the trek down the Fifteen Mile Falls was begun, and these logs were not to be purposely held back or boomed up (contained) at various places along the way. But there was no force on earth that could stop this much timber on the move. They could not stop it, they could only guide it, drive it to its destination down below.

Chapter Fourteen

White water was spraying, and men cursed as the boom and creak of great logs splashed down on Connecticut River waters below. They bumped and piled up against one another and again down below the dam they were landing like toothpicks headed in all directions and snagging the new logs that came over the dam behind them. They were headed for the "Rips" below Fitzdale, the dangerous swirling, black water, saber toothed rocks that tore at logs and bateaux and men with no more mercy for one than for the other.

Out into that water those bateaumen and those river hogs went, riding a log until its usefulness was done then stepping to another one and watching what was building up ahead. Search for the possible jam, look for the logs stuck out from a riverbank catching other logs. These were called "Jillpokes" and they deviled the rivermen.

There was a muffled cry and then the sound of boards breaking and splintering. Men were yelling and at least two were in the water. What had happened? Well, Jake couldn't see, he only knew that he heard John yell something to Peter Jay and at that very minute two great logs came thundering along and literally ripped the front of the bateau out. Jake was in water in the boat and Peter Jay was tossed out on impact. John had succeeded in grabbing a log by the middle and dragging himself upon it and he was sitting there straddled it, shaking his head to get rid of the water that was in his hair and ears and eyes.

The danger flag had been raised warning all that there was trouble up ahead, but it wasn't quite quick enough and Peter and Jake still struggled in the rock strewn water. Jake was looking around at what chance he had, if he had a chance, in this dreaded spot. He was a good swimmer but logs were shooting by him and he somehow was managing to get out of the way.

Jake thought, if anyone thinks those logs aren't going like "bats out of hell," they ought to be right smack dab in the middle of them.

Peter Jay, sat quietly on a rock that he had been able to get a hand on. Just as Peter made his first attempt to do that, a "jillpoked" log sticking out of a bank caught another one and tossed it back and Peter went feet first down into the swirling water, so as not to be run through by the thrown log. He surfaced shortly, and found that rock again and made for it once more. The threatening timber had gone down past where Peter sat and got his bearings.

The bateau was taking on water and going under and Jake jumped free just before it did. He was in a spot not too inhabited by logs and he began to tread water and made for the shore as fast as he could.

Someone threw Big John a peavey and he stood and began prying and working with it and made his way across logs until he reached the shore. Blood ran from a cut eye and on down into his left ear. The side of his head had caught a glancing blow from a log going by when the bateau first broke up. But, once on the shore he found he had a black eye to go along with the cut.

Peter Jay was bare to the waist and his saving rock had cut three long gashes in his shoulder that should have been sewn but were not. When the logs slowed down and were headed south, a bateau maneuvered its way through the raging water and plucked Peter into the boat.

"What are you doin' on that rock Jay? You're supposed

to be a bateauman!" Someone yelled this from the bank and Peter Jay gracefully got into the waiting bateau and called back,

"Old Jay, him takin' up rock climbing," and they all laughed and helped him mop up the bloodied shoulder.

Three men in the water all at once and three men out with just a few minor scratches. And for Jake, with not a mark on his body, it had been a good experience. He had faced death and he had been so busy trying to stay alive that he hadn't worried a damned bit about it. That was to an Indian a good omen. He had ridden the rapids in the ride of and for his life for the first time. And this is the very part of the Connecticut River where so many men had drowned that they were given up to those awful pork barrels. A sly smile crossed the youth's normally placid face.

By God, the second pork barrel or the third with his name on it might claim his body, but he'd gotten by the first one anyway.

"So far, so good," thought Jake.

Chapter Fifteen

The men who got drenched all went to the shore to be treated to warm tea, wrapped in heavy woolen blankets and attired in dry clothing, but the spiked boots stayed to dry on their feet. For the next few days as the drive snaked down the valley headed for the sand bars on the Seven Islands below the Fifteen Mile Falls, the men on this drive worked harder than they had at any time since this drive began on that mid- April day when the ice went out at Lake Francis.

The jamming on the Upper Coos Meadows had been of no great consequence this year. The bateau crews had only had to dynamite twice to get the drive going again, the rest of the time it had been done with cant dogs and peaveys. But from here on, the old timers said that this is when the men got separated from the boys. There would be no more bateaumen showing off by driving the boats over dams. That business all stopped at Guildhall.

They had ducked railroad bridges built on piers all the way down the river. This river had everything to snag, hold up or completely stop a log drive, dams to be sluiced past, falls, miles of rapids, dry ledges, drop offs, under currents, mills and mill owners to contend with. But most of the way this year they hadn't been stopped or slowed down by any of these obstacles. In 1882 eleven lives had been taken, and in other years railroad bridges got demolished by the thundering logs, but this year loss of life was average and the drive "hadn't killed a single bridge." What would happen now as the meadows began to narrow and the river widened? This was yet to be seen.

Three days passed and with the exception of minor log jams the Fifteen Mile Falls were allowing the timber to pass and with each foot and mile they made along that muscle straining course, the closer they came to the single most dangerous place on the river "Mulliken's Pitch." Most always the rivermen carried their bateaux around it, and many, many of those who did not, and tried to run the deadly water, wound up dead and buried in empty pork barrels.

There was a mill at McIndoes and this like Wilder and Bellows Falls was a sorting place for logs and this was exhausting and exacting work as men stood in the water and shouted out over the ongoing calamity the instructions and the direction in which logs went where, according to how they were stamped or marked. The sorter was under such pressure that he had to be relieved every hour because this job was mentally as well as physically demanding.

Sorting was one of the most necessary processes on a log drive and booms and dreaded piers were necessary in sorting timber. On the lower Connecticut River a device called a fin-boom was used. This could be one or more logs joined end to end, and strategically fastened at points along the river to "sheer" the floating timber past obstructions.

Young Jake observed the change of shift of three or four sorters and he wondered how in the world a man could ever get so familiar with the different log markings and send them in the right direction at one glance. But they could, those old timers and each year they did.

The sky had been dark and overcast, and during the night the clouds opened and forgot to shut off and there again the old timers warned the younger ones that this long and hard rain would increase the current when they reached the falls at McIndoes, Vermont.

The weeping skies cleared, and the weary rivermen rolled out of their tents not far from McIndoes Falls, and they did their utmost to stop the biggest jam they had so far, but the logs were tearing down that Long Tidal River and they could not be driven. They just kept coming, one behind the other, faster and faster until there was a pile of logs higher than any that Jake had ever seen, and these lay waiting for the rivermen just below the falls at the first narrow bend.

Back and fourth across those moving logs the squirrelmen went, prying and balancing and straining every muscle and wishing they had the eyes of an eagle, to tend out and see what was ahead, but they did not for they were only human. All they could do is ride a log as far as they could and then get to another while they searched for the danger they knew was looming up ahead.

The danger flag was raised again but the boom, that guiding fence of floating logs held together with chains, couldn't hold the logs driven forward with the ferocity of the current. And the logs and the churning white water met in a fierce struggle, and while bateaumen did their best to close in on the jam they picked and pried but could not find the Key or "King" log. And then the whole damned jam that was piled in the air over 100 feet and nearly half a mile back, suddenly hauled and it was every man for himself. Bateaux were overturned, logs rolled end on end and flew through the air like eagles landing wherever they wanted to.

As the jam gave way millions and millions of feet of snagged logs began rolling on the turbulent waters of the mighty Connecticut.

Two men had been drowned, one was crushed between two logs, and another was maimed beyond recognition. His body was found downstream later, but his decapitated head was never found.

When the bateau he was riding in this time went end-over-end Jake was thrown through the air and Peter Jay was able to drag him to shore. Jake was knocked senseless for over an hour but when he regained consciousness he knew where he was and what had happened.

He did not know any of the fatalities and he was glad that he did not. It brought back again the memory of losing Red upriver.

John was bending putting cold compresses on Jake's head, but the youth rose up and asked,

"Did Beaver get out? I saw him pitch forward off a log just before I went flying through the air!"

"Lay quiet," said Big John. "Beaver got out. We all got out."

Jake squirmed on the cot under his lithe frame. He was glad but he was sad too, for those who had not been as fortunate.

"Did they put em in empty pork barrels?" He wanted to know this and with a solemn look John nodded yes. And, as a silent tear forced its way down the youth's sweat and blood encrusted face Jake said this about his fallen brothers.

"God made no better men. They wore spiked boots."

Chapter Sixteen

Each day the intruders from the north drove millions of feet of timber further and further south and deeper into country many of these northern natives had never seen. The men were getting a little tired now, but with enthusiasm they carried on. Behind the vanguard of this drive now lay the town of East Ryegate and the "dry" Vermont town of Wells River and the "wet" New Hampshire town of Woodsville lay straight ahead and going down.

The seasoned log drivers longed for the lights of Woodsville to be seen. This like Stratford was a busy railroad town. At this depot, trains and supply wagons ran day and night. And here was a place a man could quench his thirst, get him a woman or get in a fight, because waiting ahead at Woodsville was the Ammonoosuc Drive.

The steep Nash Stream drive was part of the Ammonoosuc, and it had three lakes for headwaters, each lying above the other and it was surrounded by unusually steep mountains, Sugarloaf and Whitcomb are a couple of these. This valley was owned by O'Dell Manufacturing Company, which owned a large paper mill at Groveton, on the Upper Ammonoosuc, a few miles below the point where Nash Stream pours into that river. Like Nash Stream, Phillips Brook wound around the Kilkenny Mountains on their eastern flank and fell into the Connecticut River at Groveton.

Connecticut Valley Lumber Company's drive came from further north and sometimes the International Paper Company's drive was let out of the Ammonoosuc and actually preceded the Connecticut Valley Lumber logs to their destination at Bellows Falls. Here they were usually held up and locked securely behind a boom above the dam across the Ammonoosuc at Woodsville until the larger Connecticut Valley Lumber drive had passed. There lay the Ammonoosuc timber bumping and complaining and just plain waiting for the arrival of Van Dyke and his Rivermen. This was done annually so as to avoid trouble with the ominous and cantankerous Connecticut Valley Rivermen. It was also done because it was the sensible thing to do because if the Ammonoosuc Drive was lying in the river at Bellows falls when the CVL logs got there, it was not an easy task to sluice the bigger drive through, and they would go through. If you worked for George Van Dyke you didn't do second.

Camped on the green meadows across from Woodsville the men in the white tents were waiting for nightfall. Woodsville held the promise of good things to come. It also had a feature most of them tried to forget about, a judge and a jail. And like when the drive passed through Lancaster, New Hampshire a few days back, if you think the town was waiting with open arms, it wasn't, but the jail was.

A sly grin stole across Jake's face. Come hell or high water, come big snow or little snow the drive was going down and tonight under the glow of that old moon most of these men were going to go "visiting."

"Tom Catin' — plain and simple Tom Catin'" Beaver grumbled and with a wicked glance Big John answered simply, "MEEOOWW!"

Chapter Seventeen

Wood splinters flew in all directions from the spiked boots of the log drivers milling around the platform of the depot at Woodsville. This was known to be a railroad town, but now it was full of log drivers and there was some conflict between the railroaders and the loggers. Each were fiercely proud of their work and the swagger of the log driver was annually looked at with a bit of scorn by the trainmen.

They clomped along together Big John and Peter Jay. From one of the many saloons a piano was making some good music. Once in a while there would be a gale of hearty laughter or some woman's high pitched shriek for God only knows what reason.

John was about to reach for the clay pipe in his back pocket when he saw a woman seated in a wagon alongside the platform. She had strawberry hair and when he got closer he stopped in his tracks. It was her. It had to be her, all dressed in black with sea-green eyes. He felt a momentary twinge of guilt. His wife had been dead just over a year and he wondered if he should be attracted to or even looking at other women. Then he decided that it was alright. Hell it was just a natural thing, after all he was a widowed man, not a dead one.

"Peter, that's the woman from North Stratford! You know the one they called the witch!"

They were directly in front of her carriage now and Peter Jay looked sharply at the small female inside with the natural curls and the pretty face.

"That's her Peter, remember, they called her a witch and she said she was."

Peter Jay's face was impassive now, but he gave Sable Lore McCutcheon a second close looking over and grunted, "She is."

John's face was filled with disgust. He shook his head and was noticing for the first time the little white dog that Sable cradled in her arms. It was white as snow and some kind of a long haired mongrel.

"She's a witch?" John said sardonically and Peter only repeated "She is!"

John countered,

"Alright Jay, I suppose that little white dog in her lap is a black cat?"

"Him was," grunted Peter and then walked away leaving John alone to either stare at Sable or look straight at a sign that read Welcome to Woodsville.

John then unconsciously removed his battered felt hat and ventured forward to confront a witch and her dog or her cat or her frog or whatever in Hell witches traveled with.

"Hello," he smiled as he approached the wagon. " Do you remember me? My name is John and I met you up in North Stratford just after you'd turned a big fella into a little fella."

His easy grin was friendly and Sable did recognize him. He had escorted her to the top of the stairs while the crazy old woman kept calling her a witch! Who could forget something like that was Sable's first thought.

She smiled up at him. And when he stepped nearer to the wagon the little dog in her lap began to jump around and growl. He was as deep throated as a timber wolf, and he was a man-hater and very protective of his tiny mistress, but when he opened his pink little mouth he only had one incisor tooth. He was old. He is probably the old lady who called her a witch, thought John, but Sable met his laughing eyes with her own sea-green gaze and she answered what he was thinking,

"I found him up in Stratford. He was hungry and a drunk man kicked him in the face. That is when he lost his teeth poor little thing," said Sable.

"How did you rescue him from a mean drunk?" John questioned and Sable replied,

"I walked up to him, spit in his dirty face, and then kicked him right in the shinbone twice with my peaked toed shoe!" John laughed at the thought of the sight, and Sable continued,

"He is a scrapper that little Ragmop of mine. He got the man by the pant leg and snarled and growled and pulled until the brute lost his balance and fell as flat as a pancake. That's when I laced him right over the head with my valise so hard that the clasps broke and all my money went flying everywhere." John raised his eyebrows and Sable continued, "Well, many scampered to get it, but not too much found by the keepers was returned to the weepers, Ragmop and me. The good thing was that we were not penniless, just temporarily homeless, but we have each other now. I took the Ragmop and cleaned him up and we boarded a train headed south. We are going to Boston for a few weeks. I'm a business woman and I have ties in many places."

In his mind's eye, it was like John could see this woman and this miserable scrap of raging fur that she called a dog, fighting with a hateful drunk. He smiled, her heart was apparently kind. If she was a witch, she was at least a compassionate one. John scratched his head. He wasn't sure about her traveling companion, the black cat that was disguised as toothless dog. It hadn't done a thing but growl, bark, and snarl at him since he tried to speak with Sable.

At that moment, Ragmop decided that John was becoming too friendly with his mistress. He may be old but he was feisty! He sprang out of Sable's arms, and down out of that wagon he pounced and the next thing Big John knew, two pounds of fur was snarling and dragging on his spiked boot.

Ragmop was looking for a pantleg but John's were tucked securely into his leather-topped boots and John thought that vicious, spitting little-killer-canine must think he was chewing on some tough damned wool. He tried to shake the dog off him, but Ragmop had other plans and with his one good fang that mean tempered little beast hung onto the leg of the leather boot like a drowning man on the river hangs onto a handy log. Big John laughed and shook his leg again but he did not attempt to kick at the determined little dog, after all he was wearing spiked boots. The dog only had one tooth and toenails. Big John could still in all tell that if that Christly little mongrel had spiked boots on he would be running up and down the length of him.

"He is about as determined as a weasel in a hen house," commented an angry Sable as she scampered down out of the wagon and retrieved the little dog and carefully removed his one good incisor tooth from the leather boot.

John watched himself being rescued. He knew he could have reached down and picked up the little dog himself, but not without purposely and with malice breaking his God damned neck.

"Ragmop stop that!" Sable commanded once more that the small dog stop his attack on John's boot but the dog wouldn't consider listening to his mistress and he tried to lunge forward and get another grip on the leg of the boot.

Now Sable's face was red with temper and she stamped her foot at the little dog and said crossly,

"Ragmop! You let go of the nice man's boot or I'll kick your damned little behind so hard you'll be wearing it around your neck for a collar!" She pointed at the feisty canine Ragmop, and the small, man-hating canine let go of the boot and climbed back up into the wagon. It took him a few minutes to do that though, because for the first time, Big John noticed that the dog's right hind paw was missing.

"What happened to his foot?" Big John asked and Sable climbed up into the wagon and answered him with a sharp glance, "How in Hell do I know . . . I'm not his mother!"

Sable smothered a smile at the look on his face and then took the horse's reins in her capable hands and just before she clucked to the horses John asked her,

"I see there is a prayer meeting later on tonight. They are getting the tent up right now. Some of the boys are going over tonight. Will you be there?" John went on, "You didn't tell me or I've forgotten your name."

Sable held the reins in check and answered him, "Sable Lore MacCutcheon, twice widowed. And your name is?"

"My name is John Janas. My wife died one year ago." John answered.

"I am sorry for your trouble and my own, but perhaps we are hard on our mates, John."

What a strange thing to say thought John, but then he asked her again,

"Will you be at the meeting tonight?"

Sable sniffed high up her nose and snapped the lash high over the head of the spirited horse and before they darted off she quipped, "If I am, I'll damned well be lost, friend John."

She was off then in a flurry of dirt and coal cinders and as he watched her go John had to admit that she was a capable teamster. He was also secretly wondering where her broom was parked. He didn't know who needed a prayer meeting worse him, her, or that detestable dog of her's with three legs, one fang and no Goddamned manners at all! Walking along, looking after her, Peter Jay sauntered up beside John.

John lit his pipe and asked his friend,

"I suppose you still think she's a God damned witch?"

And Peter Jay kept walking and answered, "She is."

Before John could argue he looked down at his boot with the hole in it from the one toothed man hater. He began to swear royally.

"God damn him. He pissed on one of my Boots!" John bellowed. Peter Jay shrugged and began to chuckle. John went on, "No wonder that puke of a little white dog got his Goddamned teeth kicked out!"

Here Peter Jay made a cat face or a reasonable facsimile thereof, and with what few fingers that log driving had allowed him to keep, he tried to curl those big fingers to look like cat claws and he hissed,

"Him not white dog John, Him black cat!"

Chapter Eighteen

This revival was something that came around every summer in the north country and many of the valley people came wherever it showed up. These country folks didn't always have a church to go to nor a spiritual leader to guide them and so prayer meetings were where many went to make peace with their maker and to get cleansed of their sins. It was a spiritual awakening, and in many cases it was with the most unlikely ones getting the fire of the spirit and getting help to get rid of the terrible hold that the Devil had on their lives.

Now the Devil, he was everywhere. He was a trickster and he knew just what to do to get a man in trouble. He would always work especially hard on the younger women to see if he could get them in a vulnerable state so that a man could take advantage of them. The log driver, had a reputation of being able to do that without any help from the Devil, but he was usually much obliged for any help in the matter that Satan could give him.

This year the Connecticut Valley log drive just happened to be in the right spot at the right time so that come nightfall the men could cross the fertile meadow and go to the camp meeting.

Death had been near them this whole trip and they had gotten drunk and done some sinning up in North Stratford.

And if they didn't tonight, it was only because they didn't drink enough to get drunk or stay sober enough to get "lucky."

Woodsville was a railroad town and there were always several clashes when tough men met tough men. But here in this place the Ammonoosuc Drive actually joined into the Connecticut River Drive, as the timber of two great valleys met. That majestic timber fought like Hell not to merge quietly, and so did the mighty men who brought her down. Now, the railroad men, themselves a rugged group of combatants, had a healthy respect for the rivermen, but didn't like them much and that was easy to read in the friction of the air, even at a much anticipated prayer meeting.

Beaver didn't think he'd go. He said that he hated to see some of them people flopping around like chickens with their heads cut off, but admitted that he did find the public confessions mighty interesting.

Jake was walking to the tent beside Peter Jay. He asked the Indian,

"Why do you go to them Peter?" Peter thought a minute and then he said,

"Well, some Son-a-Bitch, him stole my pocket watch!" Jake looked at him and then said,

"You mean, you're going to go in there and pray that you find out who took it?" Peter Jay gave the younger man a strange look and replied,

"Hell, no. Old Jay him listen. Maybe some fella stand up and scream him do it! Him roll on floor and ask God's forgiveness. God do it! Peter don't. Peter, him take watch, go back home."

"Why you goin' John?" Jake asked finally and John scratched his head and said

"I'm a man who occasionally likes to be in the company of righteous women. Most of the time log drivers meet only whores."

"Tonight John him look for a witch," Peter Jay said, and John scowled at him.

"She ain't goin'," John replied and Peter knew that this

was true unless she got lost, but being a White woman she could do that easy enough.

Jake picked up on the friction between Peter Jay and John. "Oh, you still chasin' after that little river witch, John?" He laughed.

"There ain't no such thing as a witch! Them is all old wives' tales. She is just a woman, a business woman and not one who turns white dogs into black cats. She don't weave spells because she or no other mortal can change one thing into another, so there."

As they neared the big tent they passed other lumbermen and many other men and women going in. Peter was in an arguing mood.

"Up on the Reserve back home in Mohawk country lives my Uncle and for years him devil his enemies by turning himself into . . . big Timber Wolf!"

Jake and John turned and looked at Peter, "It can't be done!" Big John insisted.

Peter said, "It can be done. This old uncle him do it!"

Through their incredible stares, Peter went on, "Uncle, him powerful spirit. Only become timber wolf to get even with his enemies, who were mean to him, lie to him. Cheat him . . . cheat everybody. Uncle him become wolf and run off their stock and kill it. Then wolf stalk houses at night until enemies don't dare come out, because him that wolf, stands waiting in moonlight, watching, listening and howling to frighten them. That's why him not go log driving with me this spring."

"Why," was the question on everyone's lips and Peter answered,

"Wrong time of year. Drive ready, Uncle not ready. Uncle, not Uncle . . . Him still Big Bad Wolf!"

"God Damn you, you're joshin' us or you're lying!" snapped Big John and the stoic Peter Jay shrugged and countered, "Indian don't josh about magic. Don't lie either. White men lie, not Indian and wolf don't lie either." Here the big Mohawk made a slight grin as he finished the debate,

"Wolf, him scare the Hell out men who lie, lie and cheat. Next year, ask Uncle."

"Next year," "snorted John, next year he won't be here on the drive."

"Him will, if him done wolfin'" insisted old Jay and John dropped the subject. This critter was impossible, a stubborn, unyielding man who believed that his uncle was at times a timber wolf. As they walked along, John thought that Peter Jay was one of the best bateaumen that he had ever seen, but he had a strange way of thinking.

"Peter, sometimes I can't understand what kind of man you are," John said out loud and

Jake answered, " He's Indian. Like me, he's Indian! That's what old Beaver says and Beaver, even if he farts a lot, he sees everything just right." And he always did.

Chapter Nineteen

The lumbermen made their way inside the big tent. It was rapidly filling with people of all shapes and sizes and from many different walks of life. John's crew took some seats up back and sat looking around and were surprised to see how many faces they recognized from the drive. Well, this was a good place to reflect and thank the good Lord that they had passed three dangerous places on the Connecticut River and had made it this far unscathed. This in itself could be called a miracle.

The good Christian women who wagged their tongues viciously about the bad, unchristian ones were there in great number and were hoping, no doubt, that the tarts and trollops around the area would be led here to beg forgiveness and repent for their sins. And they, the good women could hear the sins of the bad ones! That was the best part and yes, this was the place to be.

The "bad women" were here because they wanted to belong to the Christian Fold, but they silently hoped that they would not become as mean spirited and unforgiving as those saintly sisters of theirs, who sat fanning themselves and looking down their noses at the "stained" ones.

Now, these lumberjacks liked both these kinds of women and felt that each had a place in the grand scheme of life. The good ones were the ones a man married, and made the mother of his children. The bad ones were the ones that men comforted themselves with, when they were not ready for the ties of marriage and family. And, then too, many knew instinctively, even the "baddest" woman usually had some soft, compassionate qualities about her that appealed to the adventurous side of most men. They could be themselves with this kind. They could relax and not have to be what they were not .... gentlemen. They were the rouges of the river, who could laugh and dance and satisfy their sexual appetites with the "bad ones" and could turn their backs and walk away without the common courtesy of a simple goodbye or a backward glance. It was as common in the natural order of life, as rutting season for the mountain stag. But when it was over, it was over.

The preacher was a man with a long face and graying hair. He had deep set eyes, and they were sad ones too. He was tall and lean and he was a "Hell, fire and brimstone" kind of man who could tell by just looking at the assembly before him that they were on the road to Hell and damnation.

Under his intensive gaze some people wriggled uncomfortably in their seats. He named all the sins that would get a man or a woman in the damnation position, but he offered little hope or advice about ways to stay free of those circumstances.

There was some singing and a few testimonies of people who had "seen the light," and then there was a time set aside for people to come forward and confess their sins.

Peter Jay sat with his arms folded waiting for someone to declare that he was a miserable wretch of a sinner and that he had stolen a watch, but no one did. Peter was disappointed.

Jake silently gave thanks to God who had spared him from being dead and stored forever in a pork barrel.

Big John hoped that life would never get him as bitter toward man-kind as this preacher was. Why, that fellow didn't think too much of either men or women. If they hadn't fallen into mortal sin yet, it was only a matter of time before they would! According to the sermon being delivered men and women were lusting after each other like bitch dogs in heat and couldn't wait to do the devil's bidding and jump each other's bones on the road to Hell. He needed faith in both species. All his life, Big John had been told, "There is a little bit of bad in the best of us and a little bit of good in the worst of us." And he believed that this was true, because people were neither all bad nor all good, they were a mixed bag having some of each trait. And that is what made them interesting and different from one another and gave each person a unique personality.

John's thoughts ran toward Sable MacCutcheon. Now there was a unique personality! She was such a pretty young woman, twice married, twice widowed and about twice as fascinating as any female he had ever met. John was disappointed that Sable wouldn't come to a prayer meeting, but she'd take in a stray dog and kick a man in the leg for abusing the animal. Big John felt that if ever a creature needed to be at a religious revival she did! Yes, she should be here thought John as his eyes fell upon a small, elderly lady dressed in black. She wore a drawn-down bonnet and was clutching a black cat and she wriggled her way into the last available seat.

Jesus Christ! She turned herself into an old crone and she's here with Pissey Pants! That thought was crossing John's mind, but he never let on in case Jake or Peter Jay saw her.

Sable had said if she ever went to a revival that she would be lost, but it seemed that she was lost anyhow. There she was apparently dabbling in witchcraft and surviving halfway between a good person and a bad one, torn between good and evil and apparently with some sort of connection to the Connecticut Valley log drive. And, unbeknownst to John she did have a connection because in a poker game back at North Stratford, Sable had found herself the winner and half-owner in one of the smaller drives that right now was on its way down the log filled river.

She wouldn't meet her partner until she got to Windsor, Vt. He knew a lot about log driving and this should work out very well thought Sable. She knew a lot about money, especially the art of making more of it. Sable had decided that this was the business to be in now, not as a worker but as an owner. She could smell a good investment. She could make four or five times over what she invested in this log driving venture and none would be the wiser.

Who knew it was actually a woman who had acted as banker and financed that one part of this big drive? No one knew except the partner she was yet to meet, and the Connecticut River winding endlessly through the valley and forever flowing on. This beautiful and turbulent river went all the way to the sea at Long Island Sound, New York.

Over the years, it had carried hundreds of men, millions of logs and thousands of secrets to a watery grave. And, Sable confidently thought,

"What is one..... or two more?"

Was Sable in disguise at the prayer meeting? No she was not, and how she would have howled with laughter had she known that Big John thought she was.

At the end of a boring camp revival, Jake and John decided they would go for a snort of whiskey to ease their disappointment.

"Poorest damned Prayer Meeting I ever went to," remarked Jake.

From in back of the men came the voice of Beaver who said simply,

"I knowed t'would be. That Christly God damned preacher woulda spoiled anything!"

Chapter Twenty

There was a winding river up ahead. It curled and twisted and snaked its way on down as the drive left Woodsville. The Fifteen Mile Falls had been passed six days ago but the memory of that passage would not go away. This infamous place was called Fifteen Mile Falls but the water did not gradually fall in many places, it foamed and rushed like being driven by some unseen demon wind that was Hell bent on tearing to shreds logs and men or anything else that got in its path.

The log drivers could reflect back on that angry passage, but at the time they were too busy to think about what they were doing, trying to stay alive.

This was one of the most dangerous and deadly places on the drive. There was a rumbling warning sound to be heard before the first falls actually appeared. To go toward that sound often meant death, but toward it went all the logs, all the horses and all the men who had been employed to take the drive down.

The town of Fitzdale (Lunenburg) lay behind the drive now. They rembered it well because that was where the fifteen miles of falls began. It was common knowledge that more horses had been crippled there, more bateaus stove to pieces and more lives lost there than most rivermen liked to think about.

Now in the swift water headed towards Windsor, Vermont a logger knew that the White River was still to be contended with, and yet the knowledge that the formidable Fifteen Mile Falls had been passed was an encouraging thing to look back on.

Jake remembered that passage. He could still see the logs piled up like so many hundreds of match sticks on the sharp rocks that ran along the remainder of this gauntlet called the Fifteen Mile Falls.

It was a scene of great calamity with teamsters bellowing to frightened horses with flaring nostrils, filled with deadly fear by the thunderous sound of great logs being hurled to come crashing below on rocks and sand and raging water.

No one looked back when they started down the falls. There was no time for it. Tension was everywhere and about halfway down those falls Jake remembered thinking,

"Christ! I'm here halfway down Pork Barrel Ridge."

At many spots along this phase of the drive, the bateauman did more carrying than riding but this wasn't one. And sitting in the back of the bateau, Jake tried not to think of Red, that young Irishman who drowned. Now, here on the water being tossed about like a leaf by the thrashing, foaming current underneath him, Jake began to worry about Jake and cleared his mind of Red's demise. He kept his eyes only upon Big John and Peter Jay in another bateau who had paddled ahead and successfully made their way across the swirling water to help secure or tie a boom to contain a section of logs. Jake had one time shut his eyes when he saw a great sharp rock sticking out of the foaming water and he was sure that it would take the bow of their bateau out. Somehow they got by it. The very last minute Big John saw it and had called to warn Peter Jay just in time, and the two combined their strokes and maneuvered that small craft safely out around the rock.

"God, what great beatauman!" young Jake marveled, but he knew that John and Jay would call it "nothin' but luck." But it was not luck — it was skill.

From down below danger flags were waving. Everywhere along the shore men were running and only they seemed to know where, it seemed. But the closest ones knew where, and why as a tremendous jam was building. In the fray each log that shot down these falls added to the great massive and thundering puzzle of wood.

Someone was screaming in terror and someone was screaming in pain. There were men on the land, men on the logs and men in the water and suddenly a hand was seen for the last time, waving, or was it a fallen comrade saluting for the rest of the log drivers to carry on. The man whom the current took under was a good river man who had been unlucky. They saw no more of him and he never surfaced again but drowned quickly and quietly as the other men looked on in frozen horror as tons and tons of logs now lay stacked in all directions, seeming to be exactly above the spot they saw their friend go down.

To add to the calamity, directly to the right of Jake's bateau, on the shore a big bay horse that had been twitching stray logs back toward the river had with a shrill scream reared in the harness, his great feet striking the air as he clawed it. Harness was snapping and breaking as an experienced teamster and a harness boy both tried to calm the distraught animal, but the horse only rose up again shaking his great head and throwing the harness boy who was trying to restrain him by halter and bit through the air like a rag doll. The horse reared again, screaming and thrashing in his harness as his great nostrils flared and his eyes rolled in terror.

The teamster could not control the big workhorse and with a tremendous lunge the huge bay took off, with a great log still attached, which swished and thrashed this way and that catching on all in its path logs and brush and men.

The teamster still held the reins, running behind the animal and bellowing orders, but the whiffletree (which attached the log to the harness) broke and the log lurched to the left side and then to the extreme right catching the driver and sending him sprawling to the ground. He had dragged awhile behind the runaway horse but let go when a rein broke.

Tragedy had set in before the downed driver could regain his feet. He lay stunned and helpless in the direct path of two big logs that rolled toward where he lay and one of the logs had taken his head off as swiftly as any guillotine blade.

The second log had crushed the leg of another river-men fleeing, jamming his leg against a rock so hard that the bones could readily be seen protruding through the wet and bloody pant leg.

Jake sat in the bateau stunned and frozen with shock and fear. My God, all around him men were dying and all he could do was watch. What could he or anyone do? Death was everywhere and all around him. His stomach churned.

Some two hundred feet straight ahead of where he sat motionless, the great log jam grew and grew until it was taller than any building Jake had ever seen.

Men were running for dear life and bellowing like bulls while leaping from log to log, in an attempt to not be swept up and caught in the wrath of the jam. Pike poles were flying and hats were flying too.

It was a place of cursing and screaming and at that very moment Jake and many others here became believers. Only God could save them at a time like this. War must be like that.

Jake was distracted from his prayers by someone bellowing his name. He looked up to see Big John coming catlike across what seemed now to be a river of wood. He grabbed the boy by the shoulder yanking him out of his stupor and from the bateau. "Jake, follow me. Run like Hell!"and that young man did that with a speed he thought he was incapable of. Jake began stepping where John stepped and leaping when he reached the place where Big John leaped and he was still running swifter than he had ever run in all his life before and not looking back. And with catlike sureness both men made it safely to the other side.

There they caught their breath and stopped to see where the bateau Jake had been seated in was, but it was gone, already reduced to splinters and half submerged while all the while it was rapidly being covered with great thundering logs. And this went on until the great jam looked like a mountain of wood nearly a third of a mile high.

The fatality count began, one man drowned, one crushed to death, two with broken legs and one with a fractured pelvis.

"Jesus! I thought I was headed for a ride in a pork barrel, for sure." Jake said to John who was looking about in the pandemonium for Peter. Then he saw that scarred but grinning face,

"How you doing' Jay?" Big John asked and the Indian replied,

"So far, so good!"

John was already making his way across to where his friend was propped up against an available rock to support him. Peter Jay was bleeding from a deep cut on his forehead but he could still walk and he did so, stumbling along and leaning on big John.

Jake grimaced but set about helping a small group of men look for the arm and left foot of the man who had been crushed between two logs. Jake didn't find the two missing body parts, but he joined the log driver who did and they vomited together not far from the death scene. The stench of death was strong here but so were the sounds of living. There were survivors left and that was good. Life right at this minute was good.

As he staggered on up the riverbank heading toward a warm fire, Jake was proud that he hadn't become hysterical or worse yet, wet his pants. And to celebrate that fact Jake dropped to his knees in the damp spring grass and once more he wretched.

He had never been so sick nor so thankful to be alive after that unforgettable trip. Some day he would thank his Creator in a right and respectful manner and go to church. Jake had done that once already at that at the Prayer Meeting in Woodsville. He was past the falls now and he was beyond Woodsville and he remembered back to that fateful day, of how he rose from his knees, and washed the vomit off from his face. He did this with trembling hands in the cold, clear waters of the Connecticut River. And how lucky he felt to be alive and feel that icy water on his face. God that water was cold, but God life was good.

Jake had felt stronger then, and he had picked up a pike pole and immediately gone to help his fellow log drivers and survivors pick a log jam. No one said much as their muscles strained under the grueling task. There was little to say because the battle to stay alive was not behind this determined crew, it was ahead of them. Jake like all the rest knew not where the next battle would be, only that it was waiting for them straight ahead and going down.

Later that evening the temperatures dropped and a late spring sleet pounded the tents where the men slept. In the morning the log jam was more difficult to break because of the ice underfoot.

This was a hell of a way to make a living thought Jake and then he remembered hearing the old timers say that after Woodsville their was about eighty miles gone and about one hundred eighty six left to go.

The menacing river stretched on ahead and that made Jake think about tomorrow and all the places he wanted to see along this wild journey called log driving. And it was at that moment that Jake realized that Peter Jay had known what he was talking about when he said,

"So far, so good."

Chapter Twenty One

Pittsburg Falls lay a long ways behind the drive now. The Ammonoosuc and Androscoggin drives were following along and the vanguard of the drive was way past the threat of the Fifteen Mile Falls and the excitement back at Woodsville. Woodsville that was one that the boys would remember, a railroad town that for one short time in the spring was overrun with log drivers and usually the battles were fast and furious over who was going to lay claims to that small metropolis so alive this time of year with music, whiskey and women. And these rivermen thirsted for all three.

From the second story of the farm house turned inn, Sable had stood in the widow peering out across the meadow toward the river. From this vantage point, she had stood unnoticed and silently observed the vanguard of the log drive go rolling into Woodsville, and she had seen it depart. She with her companion dog would be soon boarding a train and heading for Windsor, Vermont. She would meet her business partner there, but now she had other business to attend to. The small canine Ragmop was home and Sable had decided to do some horse trading. It was midday when she clucked to the horses and on the board sidewalks she noticed a lot of men and women milling about already. There were ladies and gentlemen and others all about and Sable knew that she could be at home with either kind, those that were upstanding citizens and those who pretended to be. She also knew that there was the "wee bit of the trollop" in the best of young womanhood and a "wee bit of goodness" in the vilest of prostitutes. And anyone who didn't know that didn't know much about life, thought Sable who mumbled "They'd be dumber than a stump."

Sable made her way toward the livery stable, and her sharp eyes fell upon two figures ahead of her on the narrow road. One appeared to be a large bearded fellow, not too steady on his feet, and then it was that Sable noticed that the bearded man was slapping and shaking a stringy haired girl whom he yanked around violently. The mere slip of a girl didn't want to be with the drunken man, but he ignored her thin and flailing arms as she protested. The slight girl was crying and begging the bully to get away from her but he paid no heed and was literally dragging her along behind him. She dug her feet into the ground and they made strange tracks in the sand.

The outraged young widow in the buggy bid the horses to stop and tugged sharply at the reins.

"You there ruffian. Let go of that child!" He saw Sable beside him sitting high in the black buggy and he told her to "Go to Hell for a meddling bitch." Ah, he saw Sable but he did not see the buggy-whip until it took him across the face and chest. He let out a horrible oath but he relinquished his hold on the sobbing girl.

To the small, distraught woman Sable commanded sharply, "Get in the wagon. Be quick about it!"

The girl did so sniffing and sobbing as she did. The drunken bully on the roadside was touching his chest where the whip had cut the skin. He saw the blood on his fingers and he began bellowing like a bull. Sable stood up in the buggy and bellowed back, "You worthless bastard, you take one more step near this carriage and I'll cut you goddamned eyes out with this whip!" He was a terrible sight to see, swearing vengeance as he was and wounded, but he was looking into a pair of sea green eyes that flashed murderously at him. He was a big man, and he wanted to charge the wagon and upend it, but he also wanted his eyes and he knew that if he did what he wanted to, that crazy women with the buggy whip would blind him. She wanted to blind him or perhaps kill him, he could tell because she was smiling, daring him to take the step towards the buggy, but he did not, he cursed and stepped back instead. Sable was in fact disappointed that her adversary did back off. She could easily inflict wounds or kill anyone when she got into one of her rages, after all any of those MacCutcheon's that got hung for murder had done their share of murder before they swung by the neck for it. Sable once had told her governess that she wished she was a Mohawk Indian so she could cut the heart of her enemy out and eat the Goddamned thing. The governess, a gentle woman was appalled. Today however, Sable took advantage of the backward movement and turned her attention to the horses and they tore away down the winding road.

On close inspection Sable saw that the girl had a cut lip and already one pale blue eye was rapidly swelling and would be black. Her dress was torn and her complexion was ghostly white. Her thin, unkept hair was a mouse brown color and her stomach was popped out in front and Sable noticed for the first time that the girl who could be no more than fifteen was quite pregnant. Oh, great! thought Sable, it isn't bad enough I have a toothless dog to take care of, but now I've got double trouble on my hands. Jesus, she thought, why do I have to be such a foolish and compassionate idiot? There was poor money to be made picking up strays.

"A fine kettle of fish," she mumbled.

"Where are we goin' Ma'am?" The tearful passenger questioned and Sable clucked at the horses and brought them to a walk. "Somewhere to Hell out of here!" Sable said not another word until she drove the team up the driveway of the inn she was staying at. Ragmop was wagging his tail and welcomed the stranger by not even hardly barking.

Sable gave her newest stray a clean handkerchief and

told her to wait until she gave the horses to the hired hand, and then she would join her on the porch. When Sable did join her house guest, she asked "Are you married to that son of a bitch?" The girl answered by shaking her head and violently whispering "No. I ain't never been with him."

"Well, he isn't the father of the child you're carrying?" At another "No. Tain't him."

Sable sighed and said, "Good. The only good to come of this whole damned day." The girl then said as tears rolled down her cheeks, "I hate the father though. If I had a gun I'd shoot him!"

As she followed Sable upstairs Sable quipped,

"Hell, only fools kill enemies with guns! Smart people know better ways."

Chapter Twenty Two

Sable had rented a large, old house on the outskirts of town. It had sprawling rooms and a big, round tower that overlooked this wide, lower Valley of the Connecticut. It came complete with furnishings and its own private ghost, but Sable didn't know that until she and Ragmop moved in. The first night there, Sable could feel the presence of someone or something else. She soon discovered that one could not keep a lamp burning in the bell tower. After trying unsuccessfully, she decided that she would let the apparition use the tower in the evenings.

The new servant girl had been given the key to the door, but Sable walked up behind her and found her standing there like one in a trance. With eyes wide and round the younger woman said,

"I tried to unlock it, Ma'am, I did, but I heard a noise and a cold hand, cold as ice hit my hand and made me drop the key!" Sable looked at the girl with raised eyebrows, but she said nothing as she reached for the key and opened the door herself.

At the alarm in the girl's eyes, Sable replied,

"The door sticks like a lot of old doors do." She did not add what was crossing her mind, that God damned ghost that sleeps in the tower probably has his foot against it!

"Ma'am, I don't know why you're being so kind to me. Me bein' in a fix to boot."

Sable beckoned her guest to enter and then she greeted her dog who was barking his head off, but Sable quieted him down and introduced him as the love of her life Ragmop MacCutcheon.

Sable untied the strings of her bonnet and asked the girl to sit in the big overstuffed chair by the fireplace. She seated herself on the adjacent chair and introduced herself as Sable MacCutcheon and asked the girl her name.

"My name is Hope, Faith Hope." The older woman smiled and said,

"Well, that's a name to remember. Before we're done with it we'll probably both need hope and faith, Miss Hope." The girl frowned and then said, "Ms. Mac ... I can't lie to you. At Mrs. Wren's house they called me by my business name, Clementine." Sable cocked her head and said "Please tell me more." And Miss Hope, alias Clementine did. She was the oldest of fourteen children and her parents were both dead. Her siblings were scattered around everywhere and she had only met a couple of them once. She was brought up in foster homes and went to work in pubs and taverns as a serving girl. At one job as a barmaid she had fallen into the company and under the influence of an older woman, a widow she called Mrs. Wren. Sable nodded her head but said nothing and the girl continued.

"Mrs. Wren lied to me Ms. Mac. She told me that I could work for her and make a whole lot of money whoring at her house. She said that I would have nice clothes and perfume and that if I played my cards right I'd have a life of luxury." Now, the young woman was fighting tears but she went on, "Well I tried that life of whorin' for her, but I didn't get much money out of it, almost none and I got hit and slapped by the customers. I was soon trying to get out of that life and I thought the only way I could do it was if one of the men I bedded would fall in love with me and marry me, but that didn't happen. They said they loved me and I believed them. I gave my body away far more than I sold it, and then Mrs. Wren beat me because I cheated her out of her share."

"Her share of your body?" said Sable and she had the picture now. Sable got up and went to the kitchen and came back with some tea for them both. As she poured the tea into the cup the younger woman's hand shook, but Sable soothed her.

"Well, giving yourself away when you could have sold it is a trifle dumb, but it is better than giving any part of it to scum like Mrs. Wren. And besides that, on the lighter side, I've heard some of the finest wives are reformed whores."

"How far along are you?" Was Sable's next question.

"I only missed a couple of periods, but I puffed up real soon," the girl answered. Sable scanned the thin body and decided it was only noticeable because the younger woman was so flat everywhere else that it showed off her slightly rounded belly.

"Oh, Ms. Mac, do you think I could be forgiven for the sinner that I am?"

"Oh, yes, God loves sinners, and we are all sinners," the older woman said.

"Ms. Mac, what is your business? You ain't no whore or fancy madam are you?" Sable Shook her head and laughed a sardonic laugh and said, "No, Hope. I'm a business woman, twice married and twice widowed and twice as ugly as any woman you ever saw, if I'm crossed, but I'm no whore. I dabble in business of various sorts but not in the flesh business. I'm clever and love to beat the boys at their own game of gambling and cards. I only cheat in self-defense when I'm sure I'm being cheated. But, I feel I'm an honest woman for the most part and I think that you can be made an honest woman too. Would you like that?" The girl nodded.

Sable had sized up the girl's thin hair, the pale skin and the sunken eyes that looked very sad. She continued, "I'll call you Ms. Hope and Ms. Mac is fine to call me." Sable smiled to herself, "Ms. Mac" was a great name in fact, because no one would know it.

Sable drank the tea and then went on, "Both my dear departed husbands left me money and I have ventured to invest it and make more of it, after all what good is money if you can't make more of it?" Sable was silent a minute and then continued,

" Well, I will consider hiring you as my maid, to do what chores you can until the baby comes and we'll deal with that, when it happens, but right now if you would be interested in that kind of work, I'll give you a decent wage and Sundays off. I won't work you too hard, but I expect you to do maid things and to do them well and to treat me with respect, but more importantly, you must from this day on respect yourself, Hope. I can keep around a good woman who made some mistakes, but I'll not tolerate sharing my home with a two-bit whore! That would be my mistake and I don't make too many."

In her most businesslike voice Sable stated,

"I rise at six a.m. You pick your own time, but I want my breakfast cooked and served to me in my room at seven a.m. sharp. My day will end with you drawing me a hot bath when I tell you to. Is that acceptable to you?"

As Sable finished, the girl rose to her feet. She was so startled by the job offer that she dropped the cup. It landed on the thick rug but it did not break and was nearly empty of liquid.

"Oh, Ms. Mac, I'll be the best maid you ever had. You will see. I'll make you proud that you gave me this job. I'll cook and I'll clean for you. What do you want me to do first, Ms. Mac?"

"I want you to march yourself into the bathroom and take a long, hot bath and wash your hair too. I'll bring extra towels. We'll burn the clothing you have on and I'll give you some of my old things, until you have earned enough to buy some new ones. You have a new life inside you and a new life ahead of you girl, but right now you stink worse than a nanny goat, so stop sniveling and start scrubbing! And don't you forget, I hire only the best!"

"Because you are the best, Ms. Mac," the thankful new maid replied and Sable cocked her head and laughed, "You'll think so, first time you burn the God damned potatoes! And no more sniveling, I hate burned potatoes and I hate sniveling!"

Chapter Twenty Three

The vanguard of the drive made its way fearlessly down the Valley of the Connecticut. The men and the horses were showing increased signs of weariness. The White River drive had joined in and the Long River flowed on before them heading for the great wide open country of Brattleboro, Vermont. The land would grow ever more flat and it would not be long before the wide, open country of Western Massachusetts appeared before them.

Here in the lower part of this great Valley of the Connecticut, the country was all more or less open, and the river itself was very wide. These men from the north were used to the Connecticut River being deep and deadly in places, but here it was wide, a great, wide river with all the dangerous whirlpools and undertows of northern water. The older men warned the younger ones of places where the water was close to a hundred feet deep, black and waiting.

Beaver made a note that just before the drive arrived at Turners Falls in Montague, Massachusetts a pair of nesting eagles had been seen. They had perched at the top of a mountain that rose up sharply from the valley and had observed the men at work for two or three days, not seeming to be startled in the least. The drives down the Connecticut always saw their share of wildlife, deer, bear, and once in awhile a Canadian lynx. Bobcats were common but the night the camp had a visit from an eastern cougar was the most memorable of all.

Back at White River Junction, a farmer carrying a rifle had come down to the meadow to talk to the foremen. Most thought it was just another incident of a farmer being mad at the drive for monies owned him, but it wasn't. They later learned that a cougar family had been pestering and had killed some livestock in the valley. From that warning, the cooks had been more careful how and where they disposed of any leftover food. The meat was put in huge canisters and hung high in the air, just the way it had to be from wolves in the north country.

The evening of the unfriendly encounter, the horses tethered nearby were the first to be aware. They were nervous, stomping and whinnying. As night fell in the meadow, the men kept their fires going longer than usual.

Jake's trusty rifle was kept locked up and dry usually but tonight it was near him. Old Beaver was glad of that. He thought it doubtful that the cougar would come. He said so earlier that night, smoking his pipe by the campfire, but that cantankerous Indian Peter Jay had listened and argued, "Catamount him thinkin' 'bout come to see us."

"How in Hell do you know that?" someone else fired back and Peter grinned and said,

"Horses. Horses known him watching for good time to come. Horses know him will come, but only old Catamount know when." Jake had only used his rifle twice on this drive, once to scare a hungry female bear who was after the garbage and once to put down a horse that had been cut so badly by a careless ax that his left hind leg could not be saved.

"If I get a good shot at the bastard, he's dead," said Jake and Peter Jay had grinned that evil Mohawk grin of his and said, "Good. You only get one!"

Beaver thought back and wondered sometimes if that Peter Jay had lived before and was perhaps a Goddamned catamount or some other wild beast like that crazy uncle of his that thought he was a timber wolf from time to time. In retrospect, one thing old Beaver knew was that what Peter Jay had said indeed had came to pass.

Beaver knew well, that Jay said,

"Big Snow (snowflakes) him mean Little Snow, and Little Snow (snowflakes) him mean Big Snow." And old Jay he was always right about the weather but, Beaver hoped he was wrong about the cougar, even though he wouldn't bet on it.

Darkness now covered the valley. Suddenly, out of the stillness of the night, the most ungodly and throaty scream that Beaver had ever heard echoed across the sleeping meadow. The cougar was somewhere close, skulking around stalking them all. The horses were going crazy and were rearing up on hind legs striking and this time when they whinnied it was not a warning, it was in stark terror.

Half dressed men were running everywhere. Teamsters tried to quiet horses wild with fright and while Beaver's hands fumbled with the kerosine lantern Jake had pulled his pants on and with no shirt and no shoes was trying to listen and see which end of the row of tents the mountain lion had visited.

"Hold that Goddamned lantern steady Beaver, I can't see. I can't see!" And Beaver was doing his best to hold that flickering light where it would most help Jake, but he wasn't doing a very good job of it.

The big cat screamed again that blood-curdling yell and sprang through the air high over a pile of dying embers that had once been a glowing campfire. Past the row of white tents and the confusion in and around them went the cougar. One riverman was a distance from the tents and had gone to the woods to do his chores. He was running towards the tents and fell flat on his face because he hadn't gotten his pants on yet. They fell to his ankles and when he went down the cougar with a terrible snarl went over the top of him. Just then a great bull of a man, one that came from the Ammonoosuc Drive threw a cant dog through the air and it just grazed the beast, and although it hurt the big cat it didn't stop him. What stopped him was when Jake saw the great beast bounding away and Beaver held that lantern just long enough so that Jake having no time to drop to one knee and take aim, took careful standing aim and fired the rifle at arm's length. And the fatal bullet took the cougar in the side of the neck. The big cat's great head snapped back and the majestic animal turned, then snarled as blood and white froth poured from his mouth. He turned as though searching for the enemy who had inflicted this wound, and his death scream was the worst and most terrifying sound of all. He was facing Jake now and that young hunter stepped back in the fury of that gaze. He held Jake's eyes like the snake holds the timid bird, and the hunter could not press the trigger to put another bullet into the cat. The cougar readied himself for the leap at Jake but the first bullet had done its work and the cougar never made the spring, but fell dead a few feet in front the hunter.

Rivermen in different stages of dress, and one completely bare, were gathering cautiously around the lifeless cat. Jake proceeded with the others but the gun was wavering in his hands.

"Jesus Christ! I thought he was going to eat you boy," someone said over the commotion and Jake stammered, "Me too."

At first light, a group of men were assigned the task of digging a hole deep enough to bury the cougar. Peter Jay asked for the teeth and the claws, they were still highly prized among his people back home. He did however ask if young Jake wanted them.

"Nope," the hunter said, "I've been as close to them as I care to!" And Peter Jay that Iroquoian puzzle of a man had smiled and said,

"You only got one. You do damned good Jake!"

The young hunter smiled. Compliments were not handed out too often on the river, but it made Jake proud to realize that once again he had cheated death on this drive.

In his journal, smeared with tobacco juice, tea and rain spots Beaver had written,

"Jake kil't a big cougar. Jay kept claws and teeth. Drive kept goin' down."

Chapter Twenty Four

At a roadside tavern in Windsor, Vermont Sable had met her partner in the drive. He was short and balding with beady eyes like a pig, so thought our Sable. He eyed her intently and made it abundantly clear that he did not like to have a woman for a partner in anything, especially in the timber business.

"Then what say, I buy your share out, then you won't have to deal with the likes of me, nor I with you." The man put down his shot glass of gin and was shocked at what this female had said to him.

"Look lady," he retorted "what makes you think you can stick your female nose into what is clearly a man's business and not loose every dime? Why don't you sell out to me?"

"You pompous old goat," said Sable and she went on, "It is a free country and I can stick my female nose anywhere I want to. I find you rude and offensive and if you're a smart man you will sell out to me at more than what you bought into this drive for. If you don't, I'll find another way to be rid of you."

He sat then and looked squarely at Sable MacCutcheon and what he saw in her eyes was enough to scare a brave man. What in Hell was this world coming to?

He smiled and squinted his small, pig eyes and swallowing another mouthful of gin he questioned, "Are you threatening me, Madam?"

Sable met his gaze with a crooked smile of her own and answered,

"I sure as Hell am, but you convince someone of it!" How angelic she looked.

"You mean that you would kill me?" the appalled man asked and the demure lady that some called a witch replied sweetly,

"No. I'd have someone else."

Sable's partner slammed the glass back down on the table and his knuckles were white and his voice was strained,

"I don't believe you'd do murder, Madam."

"Try me!" answered Sable. Her face was no longer angelic, it was evil.

Things were not going the way Sable's new business partner had planned and he was sure by now, by the coldness in her eyes that his life was in danger every day he did not sell out to her.

It was one of the most unsettling moments in his life, and even though he realized now that he was dealing with a deranged woman, she had offered to up the price he bought into the drive for.

It also crossed the balding man's mind, where in Hell she or any other woman as young as she was could get that kind of money?

He thought a long time and then he said simply,

"In the interest of my good health, I guess that I should sell out my half of this log-driving venture to you, Madam since you are inflexible and possessed." He smiled knowingly at the widow across the table from him, and she smiled back. He continued on, his eyes never wavering from Sable's innocent face,

"I'm more of a gambler than a timber man anyhow, and who knows strange woman, you may lose every dime."

"I may and I may not, but one way or the other you are out!"

And as Sable's new personal servant sat in the wagon

outside the pub and made plans to be a better person so that her mistress Ms. Mac would be proud of her, Sable was inside making one of her the best deals ever, and she cared not how anyone felt about her, proud or not proud.

In her fine hand Sable made out the agreement and the money changed from her hand to his. Just before he made his exit the man once Sable's partner, had looked at her, risen to his feet and put his beaver hat on in front of her and said,

"Madam, you are a Pirate!"

"Oh, no kind sir, that was my Father!"

Chapter Twenty Five

One-by-one this drive had passed the little valley towns that sat perched by the side of the Connecticut River. Now there were more miles of the Connecticut River Valley lying behind them than there were miles lying ahead. Sometimes these riverside hamlets were passed as quietly as a shadow and uneventful and at other places, events occurred that would keep that name in the minds of these log drivers forever.

Down, down went this drive like so many before it, snaking its serpentine way through this valley, so green and magnificent to see. Sometimes the mountains rose along both sides as though this river would have been wider, if only the mountains would have allowed it.

The flat lands of Massachusetts were waiting up ahead and there the river widened and deepened and would be no longer narrow like to the north where it began.

From Hanover to Norwich it had been a quiet ride. There was a little slow down when some straying logs struck a railroad bridge in Windsor, Vermont but nothing too disastrous nor costly. After all, this was the Connecticut Valley Log Drive going down and if something got broken, Old George Van Dyke would take care of it, or at least promise to.

The catamount incident back at White River Junction remained with Jake for many days. In his heart, he honestly did not know if he could have pulled that damned trigger one more time even to save his life. This bothered him.

In the wee hours of the morning as the van of the drive approached Bellows Falls, Vermont, a severe storm came from the north, and clobbered that section of the valley.

The thunder rumbled above and for the first time in their lives many of the men woke up to see pink lightening. No one was quite sure how big the hailstones were that pelted the tents but everyone agreed that it was a bad storm. The lightening intensified and streaked down from the sky like a wicked, pink finger pointing at places all around where the men were camped. Right in the middle of this thunderstorm a strange wind began to howl and before it was done several of the white tents had been torn down and two blew away to parts unknown.

No one was hurt except one fellow who hadn't woke up until his tent blew down. He could be seen in the aftermath of the storm, staggering around clutching his head and claiming that the hail had damaged his brain.

"Twarn't worth too God damn' much to start with." Beaver said in his astute and knowing way.

At Putney, Vermont a deranged man had driven to the meadow and tried to hire on. He was dressed in a suit and dress shoes and suddenly in the middle of his interview, he sprang out of the wagon as spry as a cat, and began to dance around like a prize fighter fisticuffing the air and calling profanities to the "Witches" he was fighting with. He said,

"The dirty bastards had tried to get into that wagon all morning!"

The men found this highly entertaining, but the man was not hired as there was no need for witch protection on the drive.... or was there?

The spacious meadows of Brattleboro, Vermont lay ahead and there was an undercurrent of excitement here because soon this "river trail" would lead them into the state of Massachusetts. They were more than three quarters finished with this trek, but they did no celebrating because death and danger and wide, unfamiliar river lay waiting around each bend. To celebrate too soon could jinx the whole thing and have devastating or even deadly consequences.

At Brattleboro, perched in a rented carriage with her infamous dog Ragmop and her faithful maid servant Hope Faith, sat Sable. She was looking down from a high mound along with a group of other spectators, watching the van of the drive maneuver its way down the valley. This drive looked like an army of ants to her, but she knew they were not ants but ruffians. But, give the devil his due, Sable realized that only a ruffian could do what they were doing.

Through her field glasses Sable saw more of what she didn't understand this thing called log driving. As they sat there, a wagon pulled up close to them and a man got out and approached Sable's carriage.

He was tall and very dark with a neat but heavy mustache. He wore a top hat and sleek black suit, and on his hands were kid gloves, that hadn't seen a speck of dirt.

He removed his top hat and nodded to the women in the wagon, but instantly the dog Ragmop disliked him even more than most men. It took all of Sable's strength to restrain the semi-toothless hound and finally she handed him over to her maid to hold.

The stranger leaned forward and held out his hand and then spoke to Sable for the first time.

"Sable MacCutcheon, I would like to talk a little business with you."

"How do you know me?" Sable questioned him. And bending low over her slowly extended hand, he said with a knowing smile,

"I've been waiting for you, Madam. I desire to buy from you my brother's share of this log drive and yours too."

Sable cocked her head at him. What to Hell was this? Now, she realized that she knew and cared little about the timber industry and this strange business called log driving which was a part of it, but she knew that it smelled of money. She knew that, and Sable liked the smell of money.

What Sable didn't like was the fact that someone she didn't know had been waiting for her. It put her at a disadvantage and that she didn't like! She hid her displeasure and looked squarely into his dark and beady eyes and said,

"How do you know that I wish to sell?" His eyes never wavered from hers and he answered, "Because money talks. Log driving is a risky business, but it is one I know about and have been involved in for years. I beg you to consider my offer dear lady, and if you meet me at the bank tomorrow morning at ten o'clock sharp, I will pay you double your investment."

Sable sat very still but gave no inkling of what she would do. She had to think about this offer. She was in the middle of something she knew nothing about, but her intuition told her there was money to be made. She could almost taste it! But, if it was a gamble, should she take the offer and run? Yes, she really had to give this some thought.

The stranger walked away and Sable clucked to the horses. She wondered how this man knew her or worse yet, what he knew about her. She frowned. What was the connection?

Their only common ground had to be his brother, the one who called her a pirate, or was it the other one, the drunk that she had cheated and easily beaten out his share of this drive up in North Stratford, N.H. in a friendly game of cards. Well, it was his own fault really, reasoned Sable. If you've been drinking too much you should stay away from horses and cards, especially away from a poker playing woman, shrewd enough to whip your ass drunk or sober.

Sable and her little family of strays would be stopping at a general store shortly to get some groceries, but she was quiet as they did this.

"Oh, Ms. Mac, that man who wants to buy you out, did you notice how cold and black his eyes were? Why, they gave me the chills," babbled Sable's maid.

"Yes," replied Sable and she ignored the part about the chills. His presence annoyed her but he didn't badly frighten her. Fine, gentle ladies frightened easily, but she was a MacCutcheon and they were a hard breed to frighten.

In her own mind, Sable was thinking to herself that she had rather spend an evening with that creepy spirit who lived in the tower room than with the critter that just left. He reminded her of a big, black snake, not a man at all. Oh well, there was more than one way to skin a cat . . . or a snake, thought Sable Lore.

Sable's newly acquired servant interrupted the quiet ride home with this spoken thought,

"Ms. Mac, I think that man who wants to pay you double is really evil." Sable nodded her head and said,

"He won't buy me out for double my investment, but for two and a half times more."

The startled maid argued,

"Oh, Ms. Mac, those black, calculating eyes like a snake, I'll bet he won't pay it!"

The daughter of old Cud MacCutcheon answered her servant with,

"He will if he buys me out . . . or he can go to Hell in a hand basket!" And Sable meant every word. After all, one good chop from a sharp ax would put a quick end to a snake . . . even a deadly one. Sable smiled.

Chapter Twenty Six

In the meadows below Brattleboro, VT a horse kicked a teamster's helper in the head and he remained unconscious hour after hour. This happened in the morning and under the cover of a steady rain he was loaded into a wagon with a tarp and carried back to Brattleboro where a doctor could see if anything could be done.

Big John rode in the front of the wagon. He went because he could just carry the youth in his arms and there was no need for a stretcher. It saddened John, to see a young man in the prime of his life that may never regain consciousness or see any more of it.

It was a tedious trip. The doctor was not at home when they arrived, but his wife directed them to an old house about four miles south. It was completely dark now and the rain was still hammering all underneath it.

At last they made their way to the house where the doctor was said to be. John rapped sharply on the door and it was opened by Sable MacCutcheon.

Water dripped from John's felt hat. He was sure you could have wrung him out, but when he explained that he was looking for a doctor, the soft voice that bid him to enter seemed familiar to him for some reason.

Inside under the lamplight, John looked down into the

face of the one the men now joshed him about all the time and referred to as the "River Witch."

He gazed intently at her. "I know you, you're Mrs. MacCutcheon. What are you doing here?"

"I'm traveling through the Connecticut River Valley. I rent this house. My maid is very ill and the doctor just arrived to see to her."

John twisted his hat in his hands. He said,

"The drive is camped on the meadow above here. This young fellow got kicked in the head by a horse. He's been unconscious for hours."

Sable bid John to bring the injured man inside. She lit yet another kerosine lamp and shortly John arrived carrying the limp figure of a youth about eighteen in his arms, as easily as if he were a small child.

"Bring him in here, Mr..." Sable gestured to a spare bedroom down the hall and John followed and said "John. My name is John." She nodded and the youth was placed on top of the coverlets of a real bed. John thought it sad it had been such a long time since the young man's frame had rested on a decent bed and he didn't even realize it.

Sable was introduced to the driver of the wagon and asked the men to please wait in the sitting room, where she would return shortly with some tea for them.

At that moment the doctor came looking for her,

"Mrs. MacCutcheon, your servant girl is very, very ill. She has lost the child she carried, but it was in the early stages and I think I can save her, though it is always questionable if she will ever have anymore. I'm encouraged that there is no infection nor fever."

Sable took the news in stride. She had thought that good care might be what the girl needed to bring a healthy baby into this world, but apparently it was not meant to be.

The next person the physician looked in on that night was the young riverman who lay with a bandage around his head, still in a state of unconsciousness. The doctor changed the bandage and said it was too early to predict the ultimate outcome. Once in a while the boy's eyelids would flutter or his lips would move or a hand would jerk, giving hope that perhaps the youth was waking up.

One thing was for sure, the longer he lay there like he was, the worse his injury probably was.

Sable went in to check on Hope Faith. Hope was in less pain than she had been. She was awake and told Sable that she would work off what her expenses were. Sable only smiled and told the girl to just rest and be strong.

"I ain't pregnant no more Ms. Mac, the doctor said so," the girl replied and Sable said, "It happens sometimes this way. They say this is nature's way of correcting something that just isn't right."

The fragile girl looked white against the bed sheets, but she knew that Sable hated crying and sniveling so she did not.

"I didn't know what I'd be like as a mother, Ms. Mac. I used to try not to think about it. Now I don't have to worry about it." Sable ran a gentle hand across the girl's forehead and said softly, "No, you don't have to worry and you should not burden yourself with regrets. What was never meant to be, will never be."

A slight smile crossed Sable's face as Hope Faith fell into a deep, even sleep.

While the doctor hovered over the fallen river man, Sable brought cups of steaming tea and some hard, sugar cookies that Hope had tried to make earlier this day before she was overcome with the cramping that began her miscarriage.

Outside the rain came down now in what seemed like buckets. The tea was hot and strong and good, but the sugar cookies would have broken the jaws of an alligator.

"Did you make these?" John smiled at Sable.

"No," she said "my cookies are worse than these!"

They both laughed and it was a good thing to be able to do on a night like this, filled with tension as two young people fought for their lives and their health. At this point the doctor emerged from the river man's bedside. He sat down and also partook of hard cookies and tea. While this was going on Ragmop went down the hall to investigate the strange man in the normally empty spare room.

Walking up to the bed the little dog stood on his hind legs and saw the strange face and the white bandages around the head. He began to bark, that sharp annoying sound that only small dogs can make. Sable went immediately to remove him from the room followed by the doctor. Just as they arrived in the doorway the young man in the bed tried to raise his right hand like he was reaching for something. Ragmop began to bark sharply again and the riverman's eyelids flickered and his lips struggled and finally formed the words, "Bbbuster .....Buster?"

Quick as a wink the doctor bent and put Ragmop on the bed. The little dog didn't mind that however, and began to sniff the youth's still hands. Then the dog walked up the bed and began licking the boy's face and a smile fell over his face, and he opened his eyes and whispered, "Buster, I've missed you. I'm glad you're here boy." A few seconds later the young log driver's eyes focused on the little dog and his hand came to stroke its fur.

No one spoke a word, but the doctor felt this was a very good sign. The patient had not responded to human voices at all, but to the sound of the barking dog something in his brain had opened up. The little dog had broken through that wall of silence, and rekindled the riverman's spirit to wake up and once again to be a part of life.

"Well, I think the mutt has done the trick," the physician smiled and minutes later when smelling salts were administered to him, the youth opened his swollen, black eyes.

John came and talked to him. The youth knew him and said,

"John, it's you John, but how'd I get here. Was Buster here?"

Big John smiled, "No, but Buster sent a friend of his to remind you that Buster is back home waiting for you."

Sable MacCutcheon was fighting back tears. The doctor explained that he would stay a while longer and that the boy should not be moved just yet. He needed rest and the fear of blood clots was the thing to worry about now, but Ragmop should be allowed to visit the patient whenever he wished.

The doctor went back to sit by the bed of Miss Hope and then departed shortly. John's driver went out to tend the horses and the estranged Sable and John went back to sit before the fireplace and have a second cup of tea.

"I owe your dog an apology. I've hated him ever since he urinated on my spike boots," John admitted and Sable laughed so hard she choked on the tea she was sipping.

"It weren't funny, not by a damned sight!" John challenged

"Well, I didn't tell him to do it!" exclaimed Sable.

"Only because you didn't think of it," John good naturedly added and she laughed again and he was once again taken by the beauty of this woman.

"What are you doing down here?" John wanted to know but before Sable could answer he added, "I know, business. You're down here doin' more business." She thought that was a very good answer so she said nothing.

"Where did you pick up the servant and where is her husband?"

"I picked her up north of here and she has no husband. She was being manhandled in the street and I brought her home with me. I just collect strays." John grinned. She really did have a good heart . . . if she had one.

"Do you have any plans about ever settling permanently anywhere, so after this drive is over a man might be interested in coming to see you?"

"No. I've tried the marriage thing twice, and done the widow thing twice and I like being alone, going and coming where and when I want."

"I didn't ask to marry you, I said I'd like to come and see you."

Sable flushed under his remark but retaliated with,

"What for. What in Hell do we have in common, you and I?"

"Well, we're both widowed and I've done a lot of thinking about you since we met. I guess I must be attracted to you for some damn reason."

"Well that's too damn bad. I'm not attracted to you! All you jokers on that log drive think you're God's gift to the women. To Hell with what the women think, they have no mind to think anything."

"I never said anything like that," John countered and then continued,

"You're a pretty woman, but you are as mean as that Christly little dog you took in."

"How do you know anything about me?" Sable snapped, setting her tea cup down with a thud. "I know," John said. "I've known all about you since I first saw you and I wish I could get you off my mind. There is nothing soft or gentle about you, Sable! There is a coldness in you girl, and it's rock hard and mighty unattractive. Perhaps you are what they called you up north ... a witch."

Sable's eyes narrowed with rage and she pointed directly at him and hissed,

"Puff, your a pile of shit!" and she continued, "Now get your arrogant, opinionated ass... out of my house!"

John got up. This had not gone very well. He knew by now that Sable was someone or something he probably would never get to know. He reached for his hat which lay in the chair beside her and he also reached for her. He kissed her then, slowly and hard, with all the harnessed feelings of tenderness of a man falling in love with a "forbidden" woman.

Sable felt his lips upon her own and they stirred within her feelings and emotions that she never had known with any of her boyfriends or with either of her two husbands. She was surprised and she was also angry that this arrogant log driver could do that. She was mad at herself and she was mad at him also. Quick as a wink, she rose to her full height, came off the ground and gave him a slap in the face that rocked his head from side to side.

"How dare you?" she snapped and went to slap him

again, but John caught her hand and held it fast in one of his own and he said, "I dare! Now don't hit me again!" She looked up at him and could see by his face that he was very angry.

Sable stiffened her body, but he pulled her to him and his lips pressed to hers once again, this time they were angry lips, lonely lips devouring hers with a passion that would not be denied. And Sable found herself clinging to him, breathless, and for the first time in her life, drowning in the kisses of a man she could not control. The more he kissed her, the more she wanted him to. She trembled.

The passionate moment was interrupted by John's driver who said that he had the wagon ready, but John told him he was staying the night.

Sable blew out some of the flickering lamps and John reached for her, picked her up as if she was a child, and carried her up the winding stairs. He gently laid her on the bed and his kisses were in her hair and on her neck, on her heaving bosom and caressing her mouth with a fierceness, yet a tenderness that demanded surrender of body and soul.

" I want you," he whispered that and no other word and Sable wanted him. She was in his arms and on fire with the passion their kisses ignited.

Shortly before daylight, he rose and left. He had loved Sable like no other man, but he hadn't said that he loved her, not once . . . not even once. She buried her face in her pillow and wept.

The night shadows departed and over the lower part of the Connecticut River Valley a new day was dawning. Sable MacCutcheon's maid was no longer with child. Sable was.

Chapter Twenty Seven

The sight of the drive coming into Turners Falls in Montague, Massachusetts is one that any man from the northern end of the Connecticut River Valley would never forget. The river was huge and the roaring water ran swift and fierce here. The swirling, foaming waters from Green River and the Deerfield River were all emptied into the Connecticut. Every kind of hazard was here, whirlpools, undercurrents, steep banks, ledges and dams that the logs had to be sluiced through.

It was said that dead men alone could make a log-driving crew at Turners Falls.

Beaver never liked this leg of the journey. In his youth, he was an able swimmer, but he was not in his youth anymore and not far from the French King Bridge, the water was still over ninety feet deep. He grimaced, a man might as well fall into the goddamned ocean!

Seated before their tents the prior evening, the crew had seen a shooting star. It streaked across the sky with a brilliance few had seen before. Most didn't pay much attention to it but Jake did. It gave him a feeling of dread. Peter Jay said to see one meant a death was coming soon and Jake tried not to dwell on this. He wished old Jay would shut his mouth sometimes. How could that Indian, who didn't talk often, get everyone's attention and "goat" when he did? Jake wondered.

They had not long to wait. A wagon arrived in the morning and the driver told one of the men who joined in on the Androscoggin drive that his brother had been shot in a drunken brawl up river. The messenger wasn't sure why the man died, only that he did.

And down went the drive and the best men were as usual, placed in the most dangerous spots along the river. The sluicing operation alone at Turners Falls was a herculean task, but there were skilled men who did it and were great at what they did. Working in that frigid water, driving logs, a man had to be careful and alert, but some fellows didn't mind that work and they would do that as opposed to breaking up a log jam with dynamite from a bateau. The morning the drive reached the infamous Turners Falls, the men were already well on their way to the end of their journey. Hopes were high, and that exhaustion that seemed to take over the bodies of men and horses, lessened with just the thought that the end of the drive was not far now. Their goal was attainable. Spirits were uplifted.

Well, whatever in Hell the log drive did now, it would make no difference to Sable. She had met with the "Snake" at the bank up river and he had met her price. She knew that he would, even if he lost every dime, because he wanted her out of his hair, and she wanted money. She grinned to herself. She had gotten into this log driving venture to make money, and she had made money!

She wasn't sure where she would go from here, but she might take her maid and her dog and go out west. If she did, she might buy a saloon! She was a good card player and a discerning gambler, and if money could be made, she would find a way to do it . . . east or west.

Sable sat at the old desk in the tower room when the doctor arrived. He checked his two young patients, and was surprised to find them both up and about.

A couple of days later, a wagon came from the drive to take the young man back to join the crew and go on down the river. Sable was once again bent over her book work in the tower when the rough, wooden wagon rumbled up the driveway.

Was it him? No, in her heart she knew that it wasn't him. He was a bateauman, not a stretcher bearer and he had more important places to be.

Sable was paid for the young stranger's board and he got into the wagon and departed. She smiled when she saw the cow-eye looks that her maid was giving the youth, but she said nothing.

The driver had told her that the drive was somewhere in Massachusetts now, and that they had made good time this year. He had spit tobacco juice and waved goodbye.

What a grubby looking old coot, thought Sable and she thought of John then, who smelled of pitch from the forest and yes, of perspiration, it was a good, male smell. Her face flushed when she thought of this.

As Sable tried to think where her next adventure would take her, her new servant girl spoke her mind,

"Ms. Mac. I really don't like this big house, do you? I don't know how you dare stay in that tower room? I swear, something awful walks about that room. I hear it! It has brought me out of a sound sleep three times now and the man who left, he heard it also."

Sable smiled, "Ah, it's probably nothing but a poor old ghost that is lonely. We have eight rooms to rattle around in that he never goes near. He is only partial to that room with the tower and if it doesn't bother me, why should it bother you?"

"It scares me Ms. Mac. Who knows what that horrible ghost could do to two women alone!"

Sable began to laugh. She regained her composure and said,

"Hope, dear, ignorant girl that you are. No ghost can ever hurt you! Watch the living, girl, you watch the living! It's the living that will hurt you!" And, when Sable thought about the living, she thought about that arrogant log driver who had made her feel alive for the first time. Yes, she knew no ghost would ever hurt her, but Oh, how this man had.

Chapter Twenty Eight

One of the most memorable days in Jake's life was when the last logs of the Connecticut Valley Log Drive were locked securely into the restraining booms at the Mt. Tom sawmills in Massachusetts. He remembered that the whole crew had rejoiced. The men did cartwheels, stood on their heads and waited with great anticipation to stand before the paymaster and finally be paid for the job that began the fall before.

He saw one very interesting thing that year. A group of rivermen from the Bangor Tigers got into a chewing-match with some of the Connecticut Valley Lumbermen and they decided to settle it with a tug of war. The contest began, and great muscles strained and the battle lasted all day with neither side giving in. Along towards evening the two teams broke a two inch hayfork or pulley-block rope. They called it a draw and the tired combatants would take their pay then and go where they felt like going. Some would get some rum and go and sleep in the local cemeteries around Holyoke. That select few of family men would wake up in the morning and not find themselves having been "rolled" with no money to take home.

Jake decided that he had a promise to fulfill. He was going to go to the infamous abode of the Black Widow Mudget, and meet that auburn-haired girl who smelled like lilacs that Red had planned to have a fling with.

Jake went to that place of ill repute and he found that girl. She was a pretty thing and Red was right, she smelled like spring lilacs. Jake told her he was there to fulfill the wishes of a friend, and it was clear by her face that she wasn't sure what or who he was taking about, but they went for a "roll in the hay," anyway. Jake smiled to himself. He had often heard Big John laughing about something being "hottern' Irish love in haying!" And Jake had met her! His grin broadened and as he started down the stairs he met a few familiar faces coming up.

In his youthful exuberance he saluted, "This one was for you Red," and then he heard the gruff voice of old Beaver, leading a pretty mulatto girl by the hand, and he barely took notice of Jake as he passed him, but old Beaver said,

"Yeah? Well, by tunkit, Jake! This one's for me!"

On the ground floor of the Mudget establishment in the town of Holyoke, John was not celebrating with the same lust as his cohorts were. He was wondering if he would ever see the "River Witch" again, and he allowed that he probably wouldn't. Sometimes when he thought of her, he wondered if anyone ever saw the real "her" anyhow.

He drank more whiskey and entered into a game of cards that ended with him and a tall unpleasant fellow taking turns losing money to each other. The man was dark complected with a heavy mustache and beady little eyes, like a snake.

He said he was an owner in this year's drive and that didn't take Big John's attention, but when he said he had bought it from a Goddamned bitch some folks called a "witch," John sat up and took notice.

"What was her name?" John asked.

"Sable Mac . . . something or other," the other man said, and John pushed the cards in front of him.

"Her name is MacCutcheon. She's no witch, Mr. and she is no bitch, either!" The man who had entered the game only after removing his kid gloves laughed softly and then said,

"I'm sorry sir, if you're a friend or a relative of this . . . woman. I didn't mean to anger you."

"Well, you sure as Hell did!" replied John.

"Sir," the darker man retaliated, "I find your anger uncalled for. But, I tell you that woman has got a name longer than your arm, and I had her checked out!"

"You take that back," said John, rising to his feet.

"I call them as I see them," the dark eyed man said. Suddenly one of Big John's fists shot out and he hauled off and hit the darker man so hard in the eye that it knocked the eye completely out of his head. "What do you see now, you son-of-a-bitch?" John snarled and the entire downstairs of the Widow Mudget's establishment erupted into utter calamity.

Peter Jay had just noticed that one of the card players at the table he was sitting at was wearing his long-lost pocket watch. Peter asked for it back, but the fellow wearing it said that he had bought the watch. Peter said, "Give it back or show me who you bought it from." The man wearing the watch would do neither, so quick as a wink, Peter pulled a knife from his boot and pinned all the paper money to the table.

"Next be your hand," he said and the man handed him the watch. At that very moment the eye went flying by and the fight was on.

Peter Jay was swinging across the heads of the merry-makers hanging onto the chandelier and kicking everyone in the head that he went by for good measure.

"God damn you Jay, You shouldn't drink!" bellowed John as the fight escalated.

"God damn you . . . you shouldn't fight over Goddamn witch!" replied the Mohawk.

And so the battle raged. It would eventually end, some of them would get tired.

Upstairs, old Beaver heard the rucous, but he wouldn't involve himself. An old fellow like him could only do one thing at a time. Besides, what did it matter who started it or what it was about? There would be another drive next year and another damned fight at the end of it.

What happened to men at the end of a log drive? Well, who really knew?

The Connecticut River knew. It knew everything about that valley and every secret that thought it still was . . . a secret. It also knew whether Sable MacCutcheon would or would not be the wife of a Connecticut Valley bateauman. Would he be Sable's husband number three? And did the "River Witch," that controversial woman, murder husbands one and two? Some guessed she did, but the Connecticut River did not guess, it knew. It still knows, but will never say, it only goes on knowing . . . goes on flowing . . . to the sea.

The End